Puppet for Containerization

Learn about configuration management and gain complete control of your Docker containers using Puppet

Scott Coulton

BIRMINGHAM - MUMBAI

Puppet for Containerization

Copyright © 2016 Packt Publishing

All rights reserved. No part of this book may be reproduced, stored in a retrieval system, or transmitted in any form or by any means, without the prior written permission of the publisher, except in the case of brief quotations embedded in critical articles or reviews.

Every effort has been made in the preparation of this book to ensure the accuracy of the information presented. However, the information contained in this book is sold without warranty, either express or implied. Neither the author, nor Packt Publishing, and its dealers and distributors will be held liable for any damages caused or alleged to be caused directly or indirectly by this book.

Packt Publishing has endeavored to provide trademark information about all of the companies and products mentioned in this book by the appropriate use of capitals. However, Packt Publishing cannot guarantee the accuracy of this information.

First published: May 2016

Production reference: 1130516

Published by Packt Publishing Ltd.
Livery Place
35 Livery Street
Birmingham B3 2PB, UK.

ISBN 978-1-78588-328-6

www.packtpub.com

Credits

Author
Scott Coulton

Reviewer
Ajeet Singh Raina

Commissioning Editor
Sarah Crofton

Acquisition Editor
Rahul Nair

Content Development Editor
Sumeet Sawant

Technical Editor
Dhiraj Chandanshive

Copy Editor
Neha Vyas

Project Coordinator
Shweta H Birwatkar

Proofreader
Safis Editing

Indexer
Rekha Nair

Production Coordinator
Aparna Bhagat

Cover Work
Aparna Bhagat

About the Author

Scott Coulton is a solutions architect with 10 years of experience in the field of managed services and hosting space. He has extensive experience in architecture and in rolling out systems and network solutions for national and multinational companies with a wide variety of technologies including AWS, Puppet, Docker, Cisco, VMware, Microsoft, and Linux. His design strengths are in cloud computing, automation, and security space.

You can find him at `https://www.linkedin.com/in/scott-coulton-22864813`. You can find him on Twitter at `@scottcoulton` and on GitHub at `https://github.com/scotty-c`.

About the Reviewer

Ajeet Singh Raina is a Docker Captain (https://www.docker.com/community/docker-captains) as well as technical lead engineer at Dell India R&D. He has picked up a variety of skills in his career, from having worked as an IT consultant and systems administrator to system integration testing. He received a certification as a VMware Certified Professional (VCP 4.1) while he was a part of the VMQA GOS validation team at VMware and has more than 8 years of industry experience. He is currently working with Enterprise Solution Group at Dell India R&D and has solid understanding of diverse range of topics, such as IT infrastructures, systems management, system integration engineering, and quality assurance.

Ajeet has a great passion for upcoming trends and technologies. He loves contributing toward open source space through writing and blogging at http://www.collabnix.com. He is currently busy evaluating and building up containers and microservices for his organization. Last year, he reviewed *PowerCLI Cookbook, Packt Publishing*.

> This book would not have been a success without direct and indirect help from many people. Thanks to my wife and 7-year old kid for putting up with me, for all the missing family time, and for providing me with love and encouragement throughout the reviewing period. Thanks to my parents and family members for all their love, guidance, and encouragement during the tough times. Thanks to all my past and present colleagues and mentors at VMware and Dell Inc. for the insightful knowledge they shared with me.

www.PacktPub.com

eBooks, discount offers, and more

Did you know that Packt offers eBook versions of every book published, with PDF and ePub files available? You can upgrade to the eBook version at `www.PacktPub.com` and as a print book customer, you are entitled to a discount on the eBook copy. Get in touch with us at `customercare@packtpub.com` for more details.

At `www.PacktPub.com`, you can also read a collection of free technical articles, sign up for a range of free newsletters and receive exclusive discounts and offers on Packt books and eBooks.

https://www2.packtpub.com/books/subscription/packtlib

Do you need instant solutions to your IT questions? PacktLib is Packt's online digital book library. Here, you can search, access, and read Packt's entire library of books.

Why subscribe?

- Fully searchable across every book published by Packt
- Copy and paste, print, and bookmark content
- On demand and accessible via a web browser

To Halen,

Dream big and work hard. There is nothing in your life that you won't be able to achieve.

*Love,
Dad*

Table of Contents

Preface	**v**
Chapter 1: Installing Docker with Puppet	**1**
Installing Vagrant	**1**
The installation	2
VirtualBox	2
Vagrant	6
Vagrantfile	9
Welcome to the Puppet Forge	**11**
The Puppet Forge	11
Creating our puppetfile	13
Installing Docker	**15**
Setting our manifests	15
Summary	**17**
Chapter 2: Working with Docker Hub	**19**
Working with Docker Hub	**19**
An overview of Docker Hub	20
Creating a Docker Hub account	24
Exploring official images	26
Automated builds in Docker Hub	**31**
Automated builds	32
Pushing to Docker Hub	35
Working with official images	**36**
Dockerfiles	36
Docker Compose	38
Puppet manifest	41
Summary	**42**

Chapter 3: Building a Single Container Application — 43
Building a Puppet module skeleton — 43
The Puppet module generator — 44
Coding using resource declarations — 48
File structures — 49
Writing our module — 49
Running our module — 53
Coding using .erb files — 55
Writing our module with Docker Compose — 55
Docker Compose up with Puppet — 57
Summary — 59

Chapter 4: Building Multicontainer Applications — 61
Decoupling a state — 62
State versus stateless — 62
Docker_bitbucket (manifest resources) — 63
Creating our module skeleton — 63
Let's code — 64
Running our module — 67
Docker_bitbucket (Docker Compose) — 70
Let's code – take 2 — 71
Running our module – take 2 — 72
Summary — 73

Chapter 5: Configuring Service Discovery and Docker Networking — 75
Service discovery — 75
The theory — 75
The service discovery module — 78
Docker networking — 95
The prerequisites — 95
The code — 95
Summary — 98

Chapter 6: Multinode Applications — 99
The design of our solution — 99
The Consul cluster — 100
The ELK stack — 100
Putting it all together — 101
The server setup — 102
The Consul cluster — 104
The ELK stack — 113
Summary — 120

Chapter 7: Container Schedulers — 121
Docker Swarm — 121
- The Docker Swarm architecture — 122
- Coding — 122
Docker UCP — 133
- The Docker UCP architecture — 133
- Coding — 134
Kubernetes — 143
- The architecture — 143
- Coding — 144
Summary — 154

Chapter 8: Logging, Monitoring, and Recovery Techniques — 155
Logging — 155
- The solution — 156
- The code — 156
  - Logstash — 163
Monitoring — 167
- Monitoring with Consul — 168
Recovery techniques — 172
- Built-in HA — 172
Summary — 176

Chapter 9: Best Practices for the Real World — 177
Hiera — 177
- What data belongs in Hiera — 178
- Tips and tricks for Hiera — 179
The code — 179
- UCP — 180
- Kubernetes — 192
Summary — 198

Index — 199

Preface

This book teaches you how to take advantage of the new benefits of containerization systems such as Docker, Kubernetes, Docker Swarm, and Docker UCP, without losing the panoptical power of proper configuration management. You will learn to integrate your containerized applications and modules with your Puppet workflow.

What this book covers

Chapter 1, *Installing Docker with Puppet*, covers how to create a development environment with Docker using Puppet. We will look at how to install Vagrant and VirtualBox. Then, we will look at Puppet Forge and how to search for modules and their dependencies. We will briefly touch upon r10k to be our transport mechanism from the Puppet Forge to our environment. Then, we build our environment with Puppet.

Chapter 2, *Working with Docker Hub*, covers a lot about the Docker Hub ecosystem: what are official images, how automated builds work, and of course, working with images in three different ways.

Chapter 3, *Building a Single Container Application*, contains our first Puppet module to create a Docker container. In this chapter, we will look at writing rspec-puppet unit tests to make sure that our module does what it's meant to do. We will know how to map our Puppet module dependencies with our `metadata.json` and `fixtures.yml` files.

Chapter 4, *Building Multicontainer Applications*, introduces Docker Compose. We will look at the docker-compose `.yaml` file construct. We will then take that knowledge and create a Puppet template (`.erb` file) and wrap that into a module. We will also touch on the Docker Compose functionality that will let us scale containers.

Chapter 5, *Configuring Service Discovery and Docker Networking*, introduces two very important topics when working with containers. First, we will look at service discovery, what it is, why do we need it, and lastly, the different types of service discovery.

Chapter 6, *Multinode Applications*, introduces all the skills that you've learned in the book so far. We are really going to step it up a notch. In this chapter, we are going to deploy four servers, and we are going to look at how to Consul cluster. In this chapter, we are going to look at the two ways to network our containers. Firstly, using the stand host IP network, that our Consul cluster will communicate on. We will also install the **ELK** (**Elasticsearch**, **Logstash**, and **Kibana**) stack.

Chapter 7, *Container Schedulers*, covers container schedulers such as Docker Swarm and Kubernetes. Then, we will build a dev environment containing four servers, three cluster nodes, and a master. We will also build a Docker network and service discovery framework.

Chapter 8, *Logging, Monitoring, and Recovery Techniques*, will take the environment that we created in the last chapter and add monitoring, logging, and recovery techniques to it. This will make our applications robust and ready for production.

Chapter 9, *Best Practices for the Real World*, focuses more on the best practices for deploying Puppet itself within a containerized environment using all the new skills that you learned in the previous chapters. By the end of this journey, readers will be able to master Puppet and Docker and apply them in the real world.

What you need for this book

For this book we need Intel i5 or above, 8 GB of ram (16 preferable), 50 GB of free disk space, and any OS that can run Vagrant.

Who this book is for

This book is designed for system administrators who are looking to explore containerization. Intermediate experience and expertise of Puppet is presumed.

Conventions

In this book, you will find a number of text styles that distinguish between different kinds of information. Here are some examples of these styles and an explanation of their meaning.

Code words in text, database table names, folder names, filenames, file extensions, pathnames, dummy URLs, user input, and Twitter handles are shown as follows: "The other change that we have made to the `servers.yaml` file is we have added entries to the `/etc/hosts` directory."

Any command-line input or output is written as follows:

```
command: -server --client 0.0.0.0 --advertise <%= @consul_advertise %> -bootstrap-expect <%= @consul_bootstrap_expect %>
```

New terms and **important words** are shown in bold. Words that you see on the screen, for example, in menus or dialog boxes, appear in the text like this: "The next thing we need to do is click on the **Create** button."

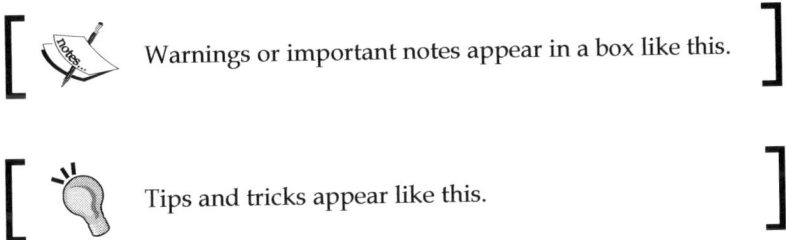

Reader feedback

Feedback from our readers is always welcome. Let us know what you think about this book—what you liked or disliked. Reader feedback is important for us as it helps us develop titles that you will really get the most out of.

To send us general feedback, simply e-mail `feedback@packtpub.com`, and mention the book's title in the subject of your message.

If there is a topic that you have expertise in and you are interested in either writing or contributing to a book, see our author guide at `www.packtpub.com/authors`.

Customer support

Now that you are the proud owner of a Packt book, we have a number of things to help you to get the most from your purchase.

Downloading the example code

You can download the example code files for this book from your account at http://www.packtpub.com. If you purchased this book elsewhere, you can visit http://www.packtpub.com/support and register to have the files e-mailed directly to you.

You can download the code files by following these steps:

1. Log in or register to our website using your e-mail address and password.
2. Hover the mouse pointer on the **SUPPORT** tab at the top.
3. Click on **Code Downloads & Errata**.
4. Enter the name of the book in the **Search** box.
5. Select the book for which you're looking to download the code files.
6. Choose from the drop-down menu where you purchased this book from.
7. Click on **Code Download**.

Once the file is downloaded, please make sure that you unzip or extract the folder using the latest version of:

- WinRAR / 7-Zip for Windows
- Zipeg / iZip / UnRarX for Mac
- 7-Zip / PeaZip for Linux

Downloading the color images of this book

We also provide you with a PDF file that has color images of the screenshots/diagrams used in this book. The color images will help you better understand the changes in the output. You can download this file from http://www.packtpub.com/sites/default/files/downloads/PuppetforContainerization_ColorImages.pdf.

Errata

Although we have taken every care to ensure the accuracy of our content, mistakes do happen. If you find a mistake in one of our books—maybe a mistake in the text or the code—we would be grateful if you could report this to us. By doing so, you can save other readers from frustration and help us improve subsequent versions of this book. If you find any errata, please report them by visiting http://www.packtpub.com/submit-errata, selecting your book, clicking on the **Errata Submission Form** link, and entering the details of your errata. Once your errata are verified, your submission will be accepted and the errata will be uploaded to our website or added to any list of existing errata under the Errata section of that title.

To view the previously submitted errata, go to https://www.packtpub.com/books/content/support and enter the name of the book in the search field. The required information will appear under the **Errata** section.

Piracy

Piracy of copyrighted material on the Internet is an ongoing problem across all media. At Packt, we take the protection of our copyright and licenses very seriously. If you come across any illegal copies of our works in any form on the Internet, please provide us with the location address or website name immediately so that we can pursue a remedy.

Please contact us at copyright@packtpub.com with a link to the suspected pirated material.

We appreciate your help in protecting our authors and our ability to bring you valuable content.

Questions

If you have a problem with any aspect of this book, you can contact us at questions@packtpub.com, and we will do our best to address the problem.

Installing Docker with Puppet

In this chapter, we will be setting up our development environment so that we can develop our first container application. To do this, we will use Vagrant. In our first topic, we will look at how to install Vagrant. We will look at how a Vagrantfile is constructed using Puppet as the provisioner. We will also look at how to get Puppet modules from the Puppet Forge using a puppetfile and r10k. In the last topic, we will install Docker on a Centos 7 box with Puppet. The following are the topics that we will cover in this chapter:

- Installing Vagrant
- An introduction to Puppet Forge
- Installing Docker

Installing Vagrant

You may ask, why are we using Vagrant for our development environment?

Vagrant is a must-have for Puppet development. The idea that you can spin up environments for development locally in minutes was a revolution in Vagrant's early releases. The product has now grown in leaps and bounds, with multiple provisioners such as Chef and Salt. Paired with multiple virtualization backends such as VirtualBox, VMware Workstation/Fusion, KVM, and we are going to use VirtualBox and Puppet as your provisioner.

The installation

Let's install Vagrant. Firstly, we will need our virtualization backend, so let's download and install VirtualBox. At the time of writing, we use VirtualBox 5.0.10 r104061. If that's outdated by the time you read this book, just grab the latest version.

You can download VirtualBox from https://www.virtualbox.org/wiki/Downloads. Choose the version for your OS, as shown in the following screenshot:

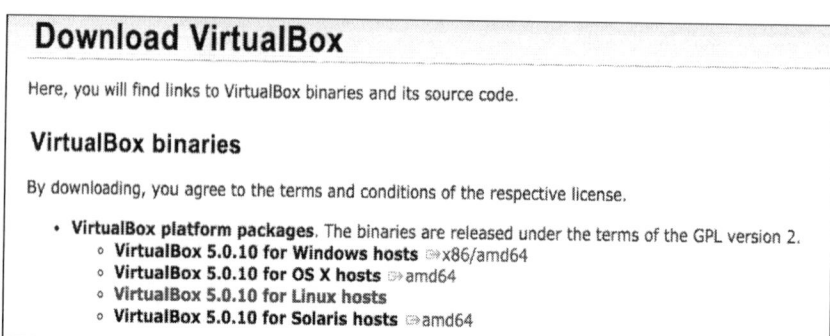

Once the package is downloaded, follow the given installation process for your OS.

VirtualBox

Follow these steps to install Vagrant on Mac OSX:

1. Go to your Downloads folder and double-click on VirtualBox.xxx.xxx.dmg. The following installation box will pop up:

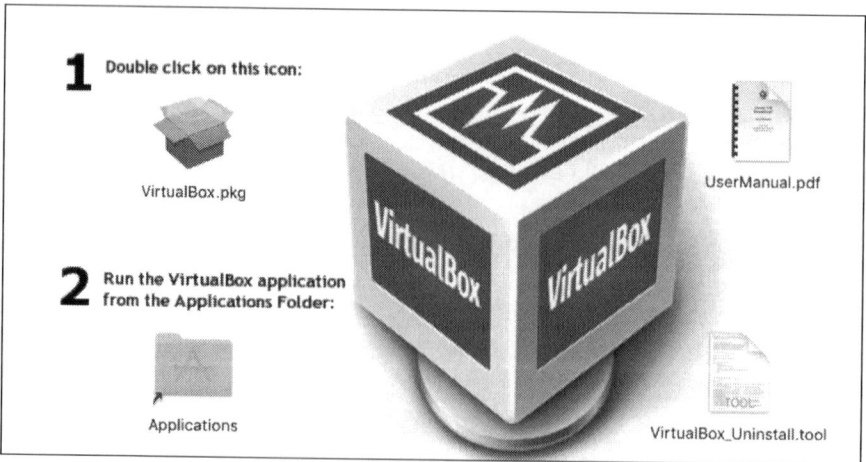

Chapter 1

2. Then, click on `VirtualBox.pkg`. Move on to the next step, as shown in the following screenshot:

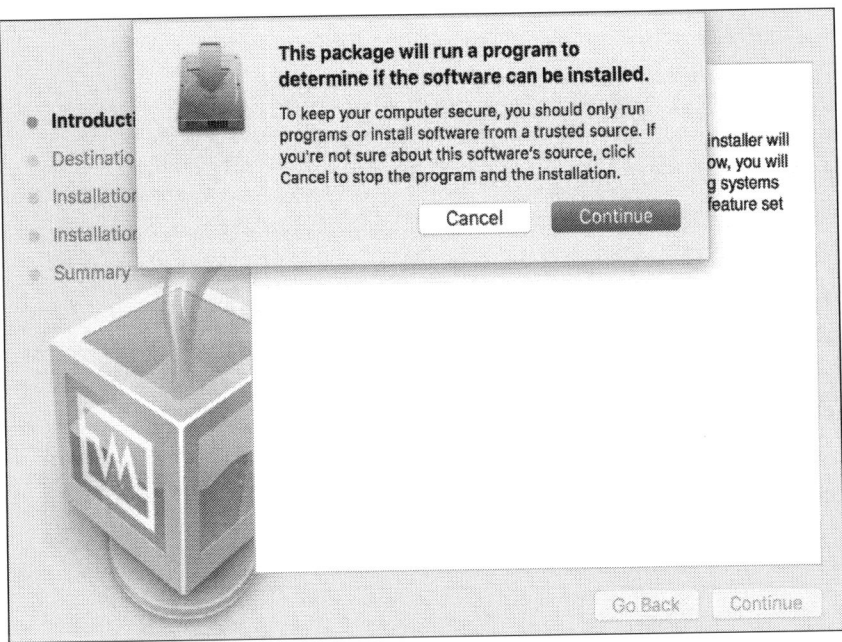

The installer will then check whether the software is compatible with the Mac OSX version.

Installing Docker with Puppet

3. After this, click on **Continue**. Once the check is successful, we can move on to the next step:

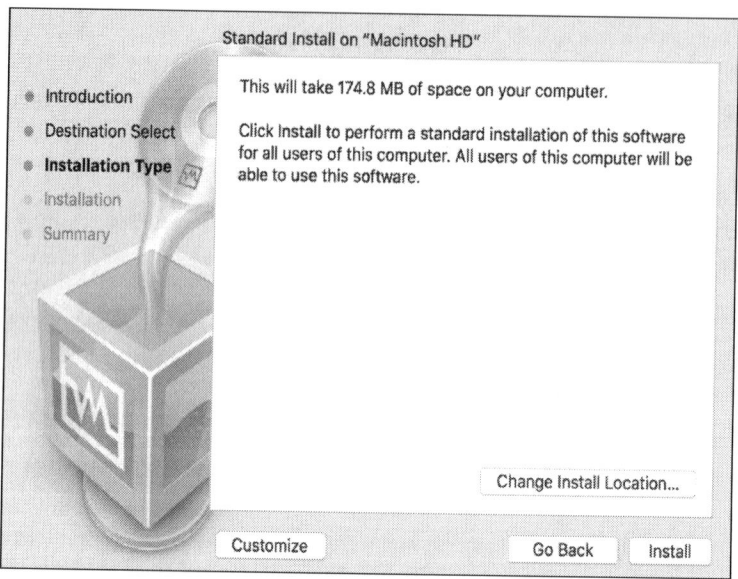

4. We then choose the default location for the installation and click on **Install**.
5. Then, enter your admin password and click on **Install Software**:

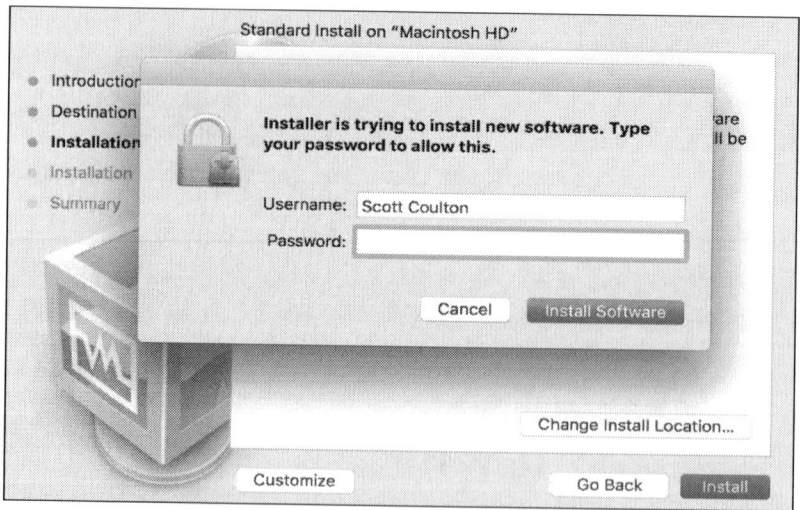

Chapter 1

The installation is now complete. The following screenshot shows what the screen looks like after completing the installation:

Now that we have the virtualization backend, we can install Vagrant:

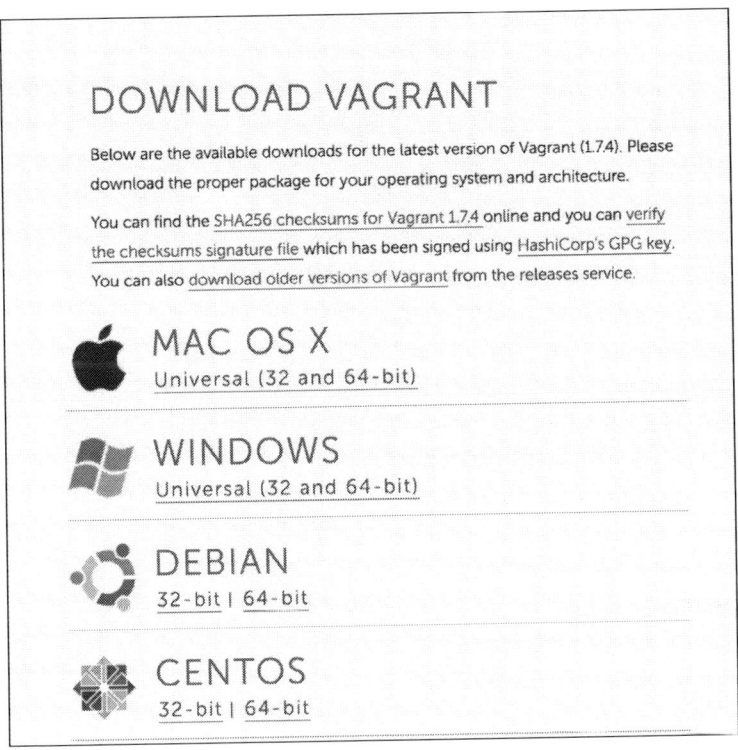

Installing Docker with Puppet

 At the time of writing this book, we are going to use Vagrant 1.7.4; if that is no longer the latest version, please grab the latest one. You can find this version of Vagrant at `https://www.vagrantup.com/downloads.html`. Again, download the installation package for your OS.

Vagrant

Here, we are just going to complete a standard installation. Follow these steps to do so:

1. Go to the folder in which you downloaded `vagrant.1.7.4.dmg` and double-click on the installer. You will then get the following pop up:

2. Double-click on `vagrant.pkg`.

3. Then, in the next dialogue box, click on **Continue**:

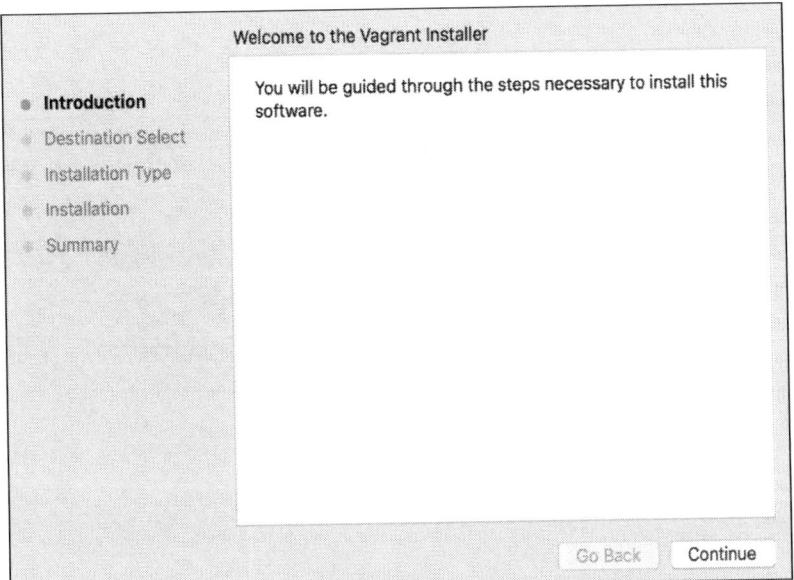

4. Then, click on the **Install** button:

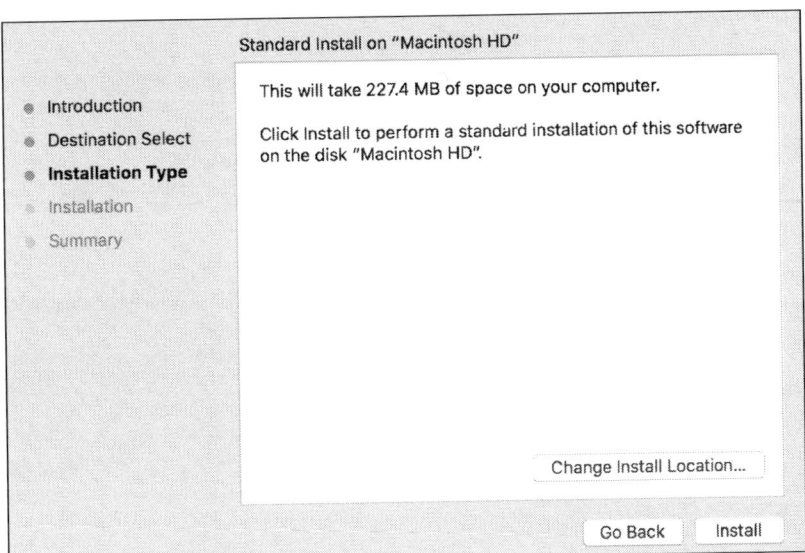

5. Enter your admin password in the given field:

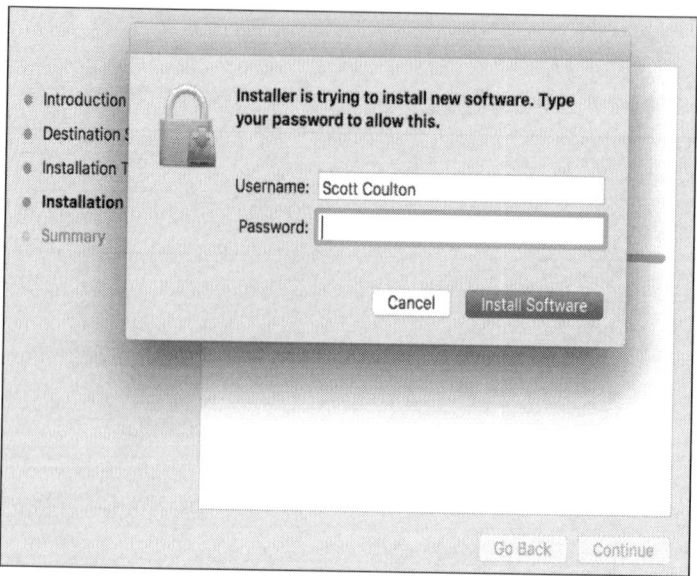

6. Once the installation is complete, open your terminal application. In the command prompt, type `vagrant`. After this, you should see the following screenshot:

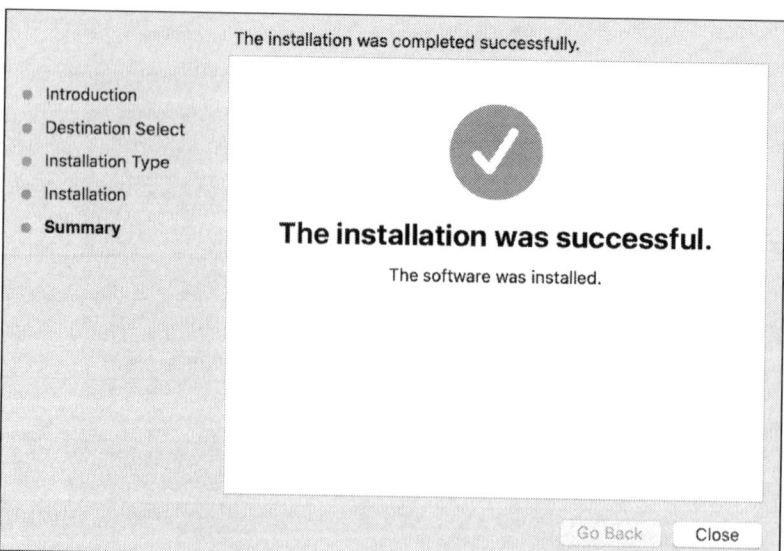

Vagrantfile

Now that we have a fully working Vagrant environment, we can start with and look at how Vagrant works and how we are going to provision our machines. As this book is not about Vagrant, we won't be writing a Vagrantfile from scratch. Instead, I have created a Vagrantfile that we will be using throughout the book:

```
✔ [scottcoulton@Scotts-MacBook-Pro] ~
08:18 $ vagrant
Usage: vagrant [options] <command> [<args>]

    -v, --version                    Print the version and exit.
    -h, --help                       Print this help.

Common commands:
     box             manages boxes: installation, removal, etc.
     connect         connect to a remotely shared Vagrant environment
     destroy         stops and deletes all traces of the vagrant machine
     global-status   outputs status Vagrant environments for this user
     halt            stops the vagrant machine
     help            shows the help for a subcommand
     hosts           Information about hostnames managed by the vagrant-hosts plugin
     init            initializes a new Vagrant environment by creating a Vagrantfile
     login           log in to HashiCorp's Atlas
     oscar
     package         packages a running vagrant environment into a box
     pe-build        Commands related to PE Installation
     plugin          manages plugins: install, uninstall, update, etc.
     provision       provisions the vagrant machine
     push            deploys code in this environment to a configured destination
     rdp             connects to machine via RDP
     reload          restarts vagrant machine, loads new Vagrantfile configuration
     resume          resume a suspended vagrant machine
     scp             copies data into a box via SCP
     share           share your Vagrant environment with anyone in the world
     ssh             connects to machine via SSH
     ssh-config      outputs OpenSSH valid configuration to connect to the machine
     status          outputs status of the vagrant machine
     suspend         suspends the machine
     up              starts and provisions the vagrant environment
     vbguest
     version         prints current and latest Vagrant version

For help on any individual command run `vagrant COMMAND -h`

Additional subcommands are available, but are either more advanced
or not commonly used. To see all subcommands, run the command
`vagrant list-commands`.
```

 You can download or Git pull the repo from https://github.com/scotty-c/vagrant-template.

[9]

Let's look at the Vagrantfile construct:

```ruby
# -*- mode: ruby -*-
# # vi: set ft=ruby :

# Specify minimum Vagrant version and Vagrant API version
Vagrant.require_version ">= 1.6.0"
VAGRANTFILE_API_VERSION = "2"

# Require YAML module
require 'yaml'

# Read YAML file with box details
servers = YAML.load_file('servers.yaml')

# Create boxes
Vagrant.configure(VAGRANTFILE_API_VERSION) do |config|

  # Iterate through entries in YAML file
  servers.each do |servers|

    config.vm.define servers["name"] do |srv|

      srv.vm.hostname = servers["name"]

      srv.vm.box = servers["box"]

      srv.vm.network "private_network", ip: servers["ip"]

      servers["forward_ports"].each do |port|
        srv.vm.network :forwarded_port, guest: port["guest"], host: port["host"]
      end

      srv.vm.provider :virtualbox do |v|
          v.cpus = servers["cpu"]
          v.memory = servers["ram"]
      end

      srv.vm.synced_folder "./", "/home/vagrant/#{servers['name']}"

      servers["shell_commands"].each do |sh|
        srv.vm.provision "shell", inline: sh["shell"]
      end

      srv.vm.provision :puppet do |puppet|
          puppet.temp_dir = "/tmp"
          puppet.options = ['--modulepath=/tmp/modules', '--verbose']
          puppet.hiera_config_path = "hiera.yaml"

      end
    end
  end
end
```

As you can see from the preceding screenshot, the Vagrantfile is actually a Ruby file. As it is Ruby, it opens up a world of opportunities for us to make our code elegant and efficient. So, in this Vagrantfile, we have extracted all the low-level configurations and replaced them with a few parameters. Why are we doing this? The reason is to split up our logic from our configuration and also iterate our configuration in order to stop replication of our code. So, where is all the configuration stored? The answer is in the `servers.yaml` file. This is where we set the vagrant box that we want to deploy, the number of CPUs for the box, the internal network's IP, the hostname, the forwarded ports between the guest and host, and the RAM and shell provider for bash commands that we need to get the environment ready for Puppet to run, for example, downloading modules and their dependencies from the Puppet Forge:

```
box: puppetlabs/centos-7.0-64-puppet-enterprise
cpu: 1
ip: "172.17.8.101"
name: node-01
forward_ports:
    - { guest: 80, host: 8080 }
ram: 2048
shell_commands:
    - { shell: 'yum install -y wget git lvm2 device-mapper-libs' }
    - { shell: '/opt/puppet/bin/gem install r10k && ln -s /opt/puppet/bin/r10k /usr/bin/r10k || true' }
    - { shell: 'cp /home/vagrant/node-01/Puppetfile /tmp && cd /tmp && r10k puppetfile install --verbose' }
```

The benefit of this approach is also that any developer using a Vagrantfile does not need to actually modify the logic in the Vagrantfile. They only need to update the configuration in `servers.yaml`. As we go through the book, we will work with the other files in the repository, such as `Puppetfile`, `hieradata`, and `manifests`. Now that we have set up our Vagrant environment, let's look at how to get our Puppet modules from the Puppet Forge.

Welcome to the Puppet Forge

In this topic, we will look at how to find modules from the Puppet Forge. Then, we will see how to pull them with their dependencies using a puppetfile and r10k. This will set us up for our last topic, *Installing Docker with Puppet*.

The Puppet Forge

One of the great things about puppetlabs and their products is the community. If you ever get a chance to attend PuppetConf or a Puppet Camp, depending on where you live, I would really recommend you to attend it. There will be a wealth of knowledge there and you will meet some really great people.

Installing Docker with Puppet

The Puppet Forge is a website that puppetlabs runs. It is a place where other Puppet developers publish modules that are ready to use. You might be asking, what about GitHub? Can't you get modules from there? Yes, you can. The difference between the Puppet Forge and GitHub is that the Puppet Forge is the stable, releasable version of the module, whereas GitHub is the place to contribute to the module, that is, a place to create pull requests.

 You can find the Puppet Forge at `https://forge.puppetlabs.com/`.

The following screenshot shows the home page of Puppet Forge:

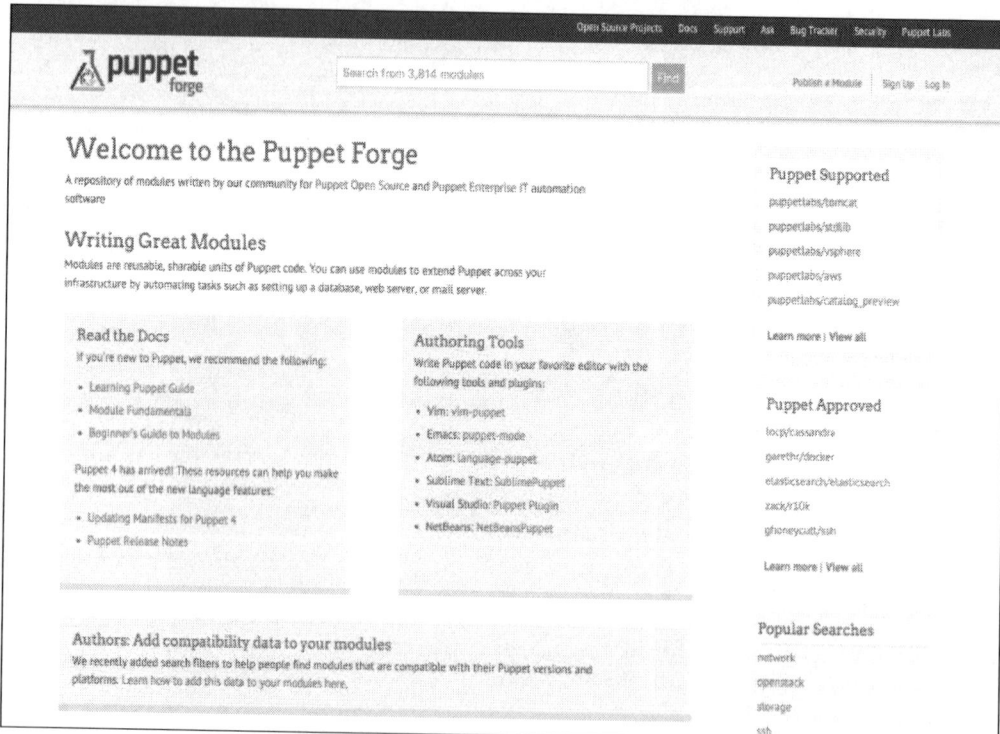

Now that we have been introduced to the Puppet Forge, let's use it to find our Docker module that we will be using to build our environment.

 We are going to use the garethr/docker Docker module, which you can find at `https://forge.puppetlabs.com/garethr/docker`.

[12]

Now that we have selected our module, we can move on to setting up our puppetfile:

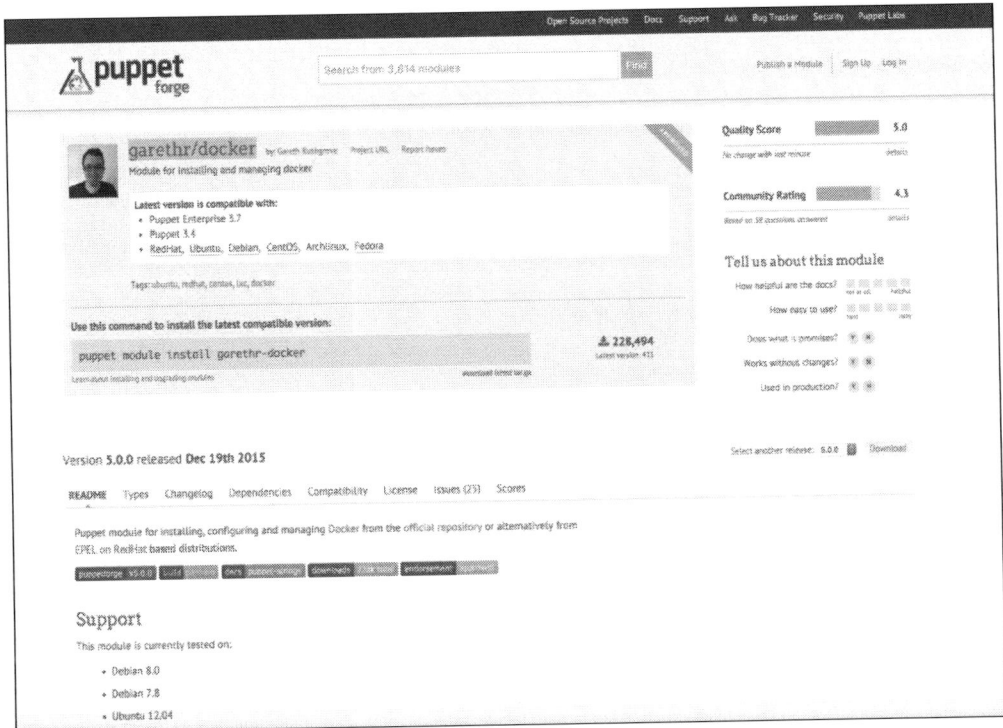

Creating our puppetfile

In the previous topic, we cloned our Vagrant template using Git. In that repo, there is also a puppetfile. A puppetfile is used as a control file for our modules. It will list all the modules that we need (in this instance, just to install Docker). r10k will then reference the puppetfile and pull the modules from the Puppet Forge into our environment's directory.

As modules have dependencies, we need to make sure that we capture them in our puppetfile. For the Docker module, we have three dependencies: **puppetlabs/stdlib** (>= 4.1.0), **puppetlabs/apt** (>= 1.8.0 <= 3.0.0), and **stahnma/epel** (>= 0.0.6), as shown in the following screenshot.

Now, we know all the modules that we need to build a Docker environment. We just need to add them to our puppetfile.

Installing Docker with Puppet

The following screenshot is an example of what the puppetfile should look like:

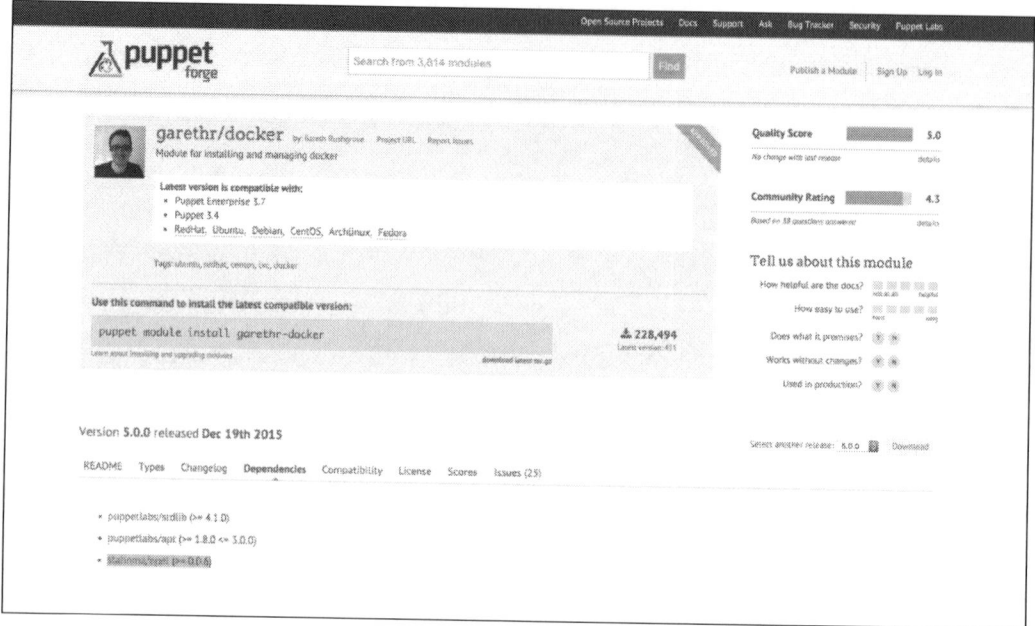

Now, when we run `vagrant up`, r10k will pull the modules from the Puppet Forge. We invoke r10k on line 13 of `servers.yaml` with the `r10k puppetfile install--verbose` command. The following screenshot shows the output of this command:

If we are successful, the terminal will provide the following output:

Now that we have our puppetfile set up, we can install Docker.

[14]

Installing Docker

In this topic, we will put together all the configuration from our Vagrant repo and knowledge of the Puppet Forge to create the Docker environment.

Setting our manifests

The first thing that we need to do to install Docker is set our manifest to include the Docker class on our node. To do this, let's go to our Vagrant repo. In the repo, there is a file in the `manifests` directory called `default.pp`. We need to edit the file to include the Docker class `node 'node-01' { include docker}`. We can now save the file, and we are ready to run our environment.

The first step is to open our terminal and change to root of the Vagrant repo. Then, we need to enter the `vagrant up` command:

```
✓ [scottcoulton@Scotts-MacBook-Pro] ~/Documents/Local Dev/Vagrant Builds/My book [master|✚ 3…1]
14:50 $ vagrant up
```

This will now provide us with our CentOS 7 box. Install r10k and then run Puppet and apply the Docker class. This will take about 4 minutes depending on your laptop and network connection. If the box was provisioned successfully, you will see the following output:

```
==> node-01: Running Puppet with default.pp...
==> node-01: Info: Loading facts
==> node-01: Info: Loading facts
==> node-01: Info: Loading facts
==> node-01: Notice: Compiled catalog for localhost in environment production in 0.99 seconds
==> node-01: Info: Applying configuration version '1450756659'
==> node-01: Notice: /Stage[main]/Docker::Repos/Yumrepo[docker]/ensure: created
==> node-01: Info: changing mode of /etc/yum.repos.d/docker.repo from 600 to 644
==> node-01: Notice: /Stage[main]/Docker::Install/Package[docker]/ensure: created
==> node-01: Notice: /Stage[main]/Docker::Service/File[/etc/sysconfig/docker-storage-setup]/ensure: created
==> node-01: Info: /Stage[main]/Docker::Service/File[/etc/sysconfig/docker-storage-setup]: Scheduling refresh of Service[docker]
==> node-01: Notice: /Stage[main]/Docker::Service/File[/etc/systemd/system/docker.service.d]/ensure: created
==> node-01: Notice: /Stage[main]/Docker::Service/File[/etc/systemd/system/docker.service.d/service-overrides.conf]/ensure: created
==> node-01: Info: /Stage[main]/Docker::Service/File[/etc/systemd/system/docker.service.d/service-overrides.conf]: Scheduling refresh of Exec[docker-systemd-reload]
==> node-01: Notice: /Stage[main]/Docker::Service/Exec[docker-systemd-reload]: Triggered 'refresh' from 1 events
==> node-01: Notice: /Stage[main]/Docker::Service/File[/etc/sysconfig/docker]/ensure: created
==> node-01: Info: /Stage[main]/Docker::Service/File[/etc/sysconfig/docker]: Scheduling refresh of Service[docker]
==> node-01: Notice: /Stage[main]/Docker::Service/File[/etc/sysconfig/docker-storage]/ensure: created
==> node-01: Info: /Stage[main]/Docker::Service/File[/etc/sysconfig/docker-storage]: Scheduling refresh of Service[docker]
==> node-01: Notice: /Stage[main]/Docker::Service/Service[docker]/ensure: ensure changed 'stopped' to 'running'
==> node-01: Info: /Stage[main]/Docker::Service/Service[docker]: Unscheduling refresh on Service[docker]
==> node-01: Notice: Finished catalog run in 22.35 seconds
```

We can also verify that the Docker installation was successful by logging in to the box via SSH. We will do that with the `vagrant ssh` command. Once we are in, we will sudo up to root (`sudo -i`). Now, let's just check whether Docker is installed with the `docker` command.

Installing Docker with Puppet

You will see the following output on the terminal:

```
[root@node-01 ~]# docker
Usage: docker [OPTIONS] COMMAND [arg...]
       docker daemon [ --help | ... ]
       docker [ --help | -v | --version ]

A self-sufficient runtime for containers.

Options:

  --config=~/.docker              Location of client config files
  -D, --debug=false               Enable debug mode
  --disable-legacy-registry=false Do not contact legacy registries
  -H, --host=[]                   Daemon socket(s) to connect to
  -h, --help=false                Print usage
  -l, --log-level=info            Set the logging level
  --tls=false                     Use TLS; implied by --tlsverify
  --tlscacert=~/.docker/ca.pem    Trust certs signed only by this CA
  --tlscert=~/.docker/cert.pem    Path to TLS certificate file
  --tlskey=~/.docker/key.pem      Path to TLS key file
  --tlsverify=false               Use TLS and verify the remote
  -v, --version=false             Print version information and quit

Commands:
    attach    Attach to a running container
    build     Build an image from a Dockerfile
    commit    Create a new image from a container's changes
    cp        Copy files/folders between a container and the local filesystem
    create    Create a new container
    diff      Inspect changes on a container's filesystem
    events    Get real time events from the server
    exec      Run a command in a running container
    export    Export a container's filesystem as a tar archive
    history   Show the history of an image
    images    List images
    import    Import the contents from a tarball to create a filesystem image
    info      Display system-wide information
    inspect   Return low-level information on a container or image
    kill      Kill a running container
    load      Load an image from a tar archive or STDIN
    login     Register or log in to a Docker registry
    logout    Log out from a Docker registry
    logs      Fetch the logs of a container
    network   Manage Docker networks
    pause     Pause all processes within a container
    port      List port mappings or a specific mapping for the CONTAINER
    ps        List containers
    pull      Pull an image or a repository from a registry
    push      Push an image or a repository to a registry
    rename    Rename a container
    restart   Restart a container
    rm        Remove one or more containers
    rmi       Remove one or more images
    run       Run a command in a new container
    save      Save an image(s) to a tar archive
    search    Search the Docker Hub for images
    start     Start one or more stopped containers
    stats     Display a live stream of container(s) resource usage statistics
    stop      Stop a running container
    tag       Tag an image into a repository
    top       Display the running processes of a container
    unpause   Unpause all processes within a container
    version   Show the Docker version information
    volume    Manage Docker volumes
    wait      Block until a container stops, then print its exit code

Run 'docker COMMAND --help' for more information on a command.
[root@node-01 ~]#
```

Summary

In this chapter, we covered how to create a development environment with Docker using Puppet. We looked at how to install Vagrant and VirtualBox. Then, we looked at the Puppet Forge, how to search for modules and their dependencies. We then took the dependencies and mapped them to a puppetfile. We briefly touched on r10k, which is our transport mechanism from the Puppet Forge to our environment. Then, we built our environment with Puppet.

In the next chapter, we'll take a look at how to access Docker Hub and pull public images.

2
Working with Docker Hub

In this chapter, we will look at Docker Hub, what it is, how to sign up for an account, how to pull an image, how to push an image, and automated image builds. This will give us a good solid foundation for future topics when we need to work with official images.

In this chapter, we will cover the following topics:

- What is Docker Hub?
- Automated builds
- Working with official images

Working with Docker Hub

In this section, we will discuss Docker Hub, what it is used for, what features does it provide, and lastly, how is it different from other repository sites, such as GitHub or Puppet Forge. We will then create an account and explore our account settings. After this, we will look at official images to get a solid foundation for the next topic.

Working with Docker Hub

An overview of Docker Hub

In the last chapter, we looked at the Puppet's repo service, The Forge as it is called by the community (`https://forge.puppetlabs.com/`). Now, let's look at Docker's repo service, Docker Hub. We can find Docker Hub at `https://hub.docker.com/`.

The following screenshot shows what the screen looks like:

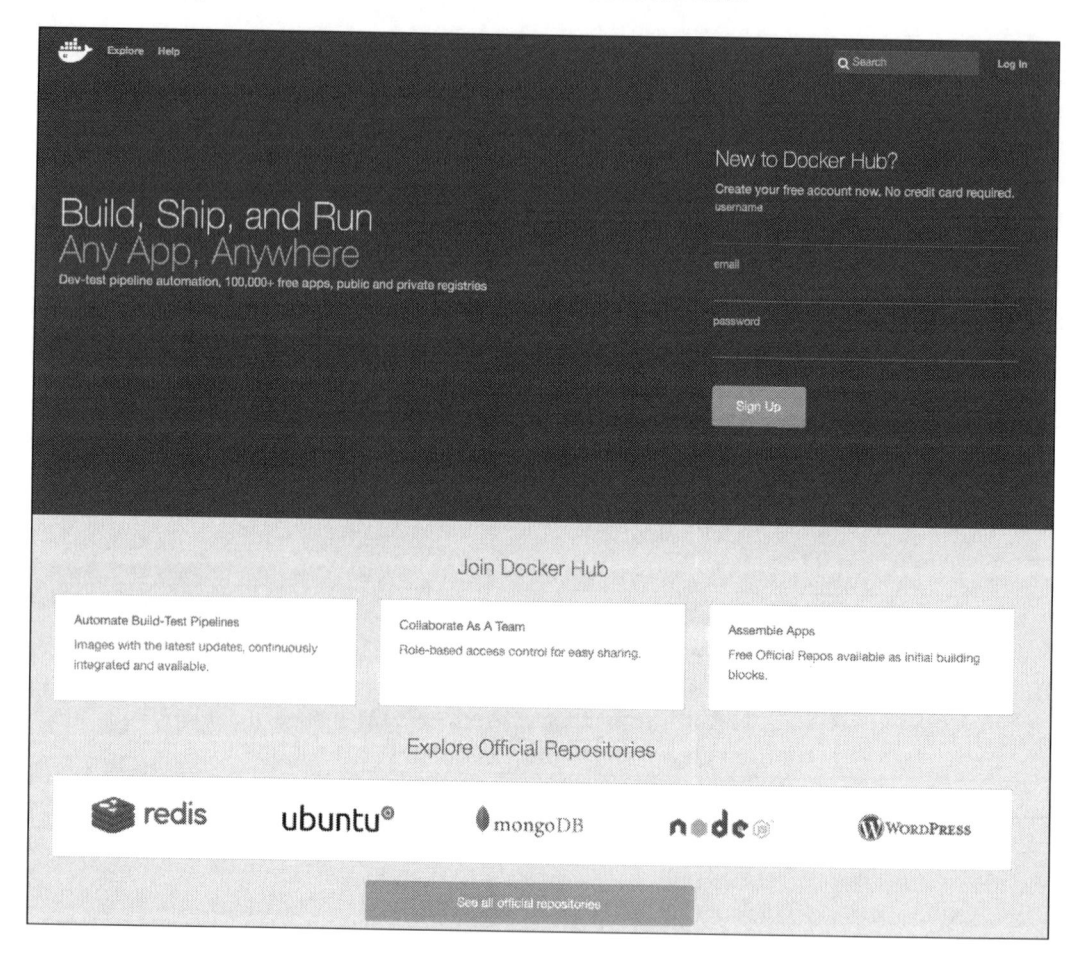

In Docker Hub, there are two type of images:

- Official images
- Images authored by developers

First, we will talk about official images. On Docker Hub, you can get official images for just about any major operating system or application. So, the benefit for you as a developer is that the work to install the application is done for you, saving you the time and effort. This allows you to focus your time on developing. Let's look at an example—we will use golang.

First, we will search for golang in the search box at the top right-hand side of the front page, as shown in the following screenshot:

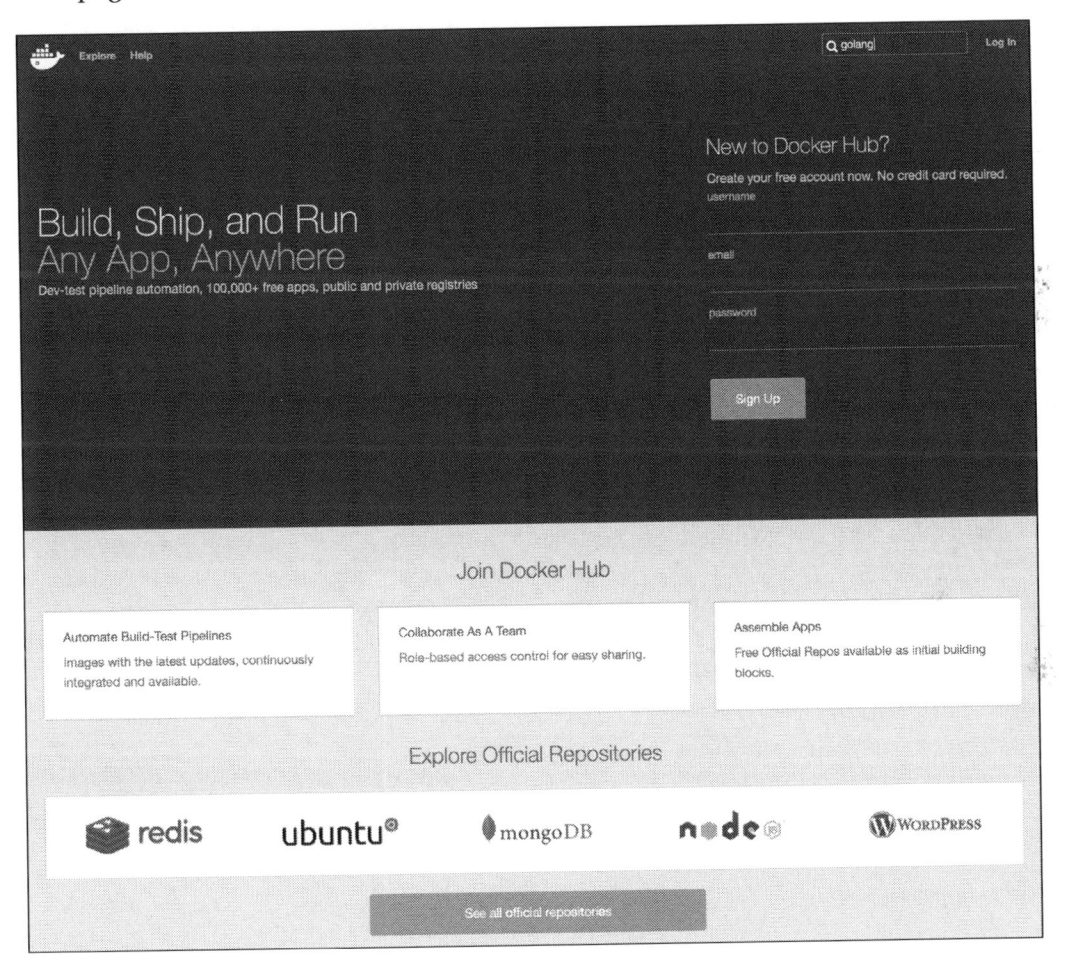

Working with Docker Hub

Our search will return the following results:

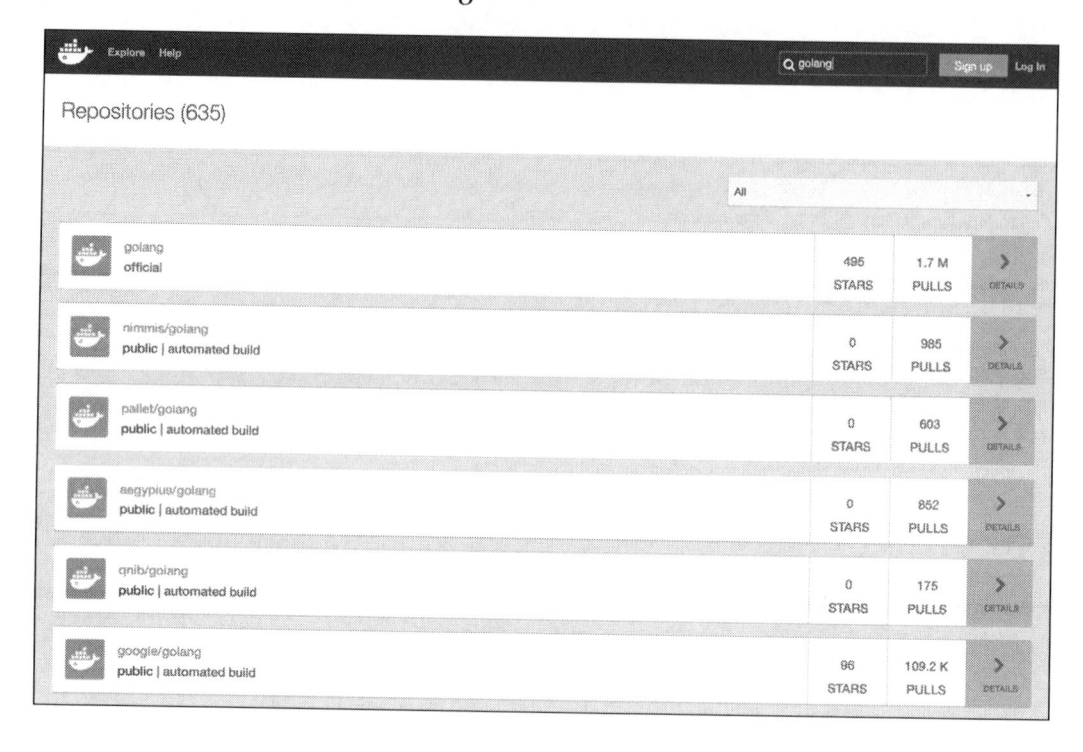

golang official	495 STARS	1.7 M PULLS	DETAILS
nimmis/golang public	automated build	0 STARS	985 PULLS	DETAILS
pallet/golang public	automated build	0 STARS	603 PULLS	DETAILS
aegypius/golang public	automated build	0 STARS	852 PULLS	DETAILS
qnib/golang public	automated build	0 STARS	175 PULLS	DETAILS
google/golang public	automated build	96 STARS	109.2 K PULLS	DETAILS

We will click on the official release of golang, as shown in the following screenshot:

Chapter 2

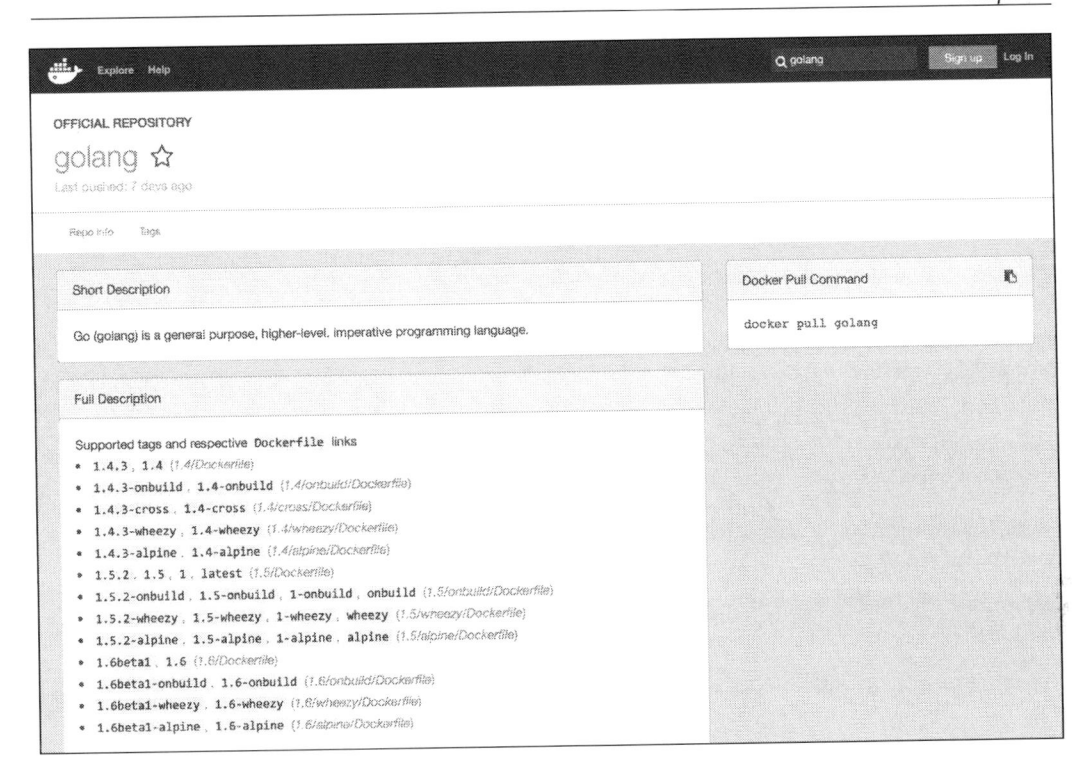

As we can see in the preceding screenshot, this repository gives us a lot of options. So, we can use multiple, different versions of golang, even on multiple different operating systems. So, to build a golang app, all we need to do is choose an image. We will use the following image in our Dockerfile:

```
1.4.3-wheezy, 1.4-wheezy
```

We will then use the `COPY` method in our Dockerfile to get our code into the container on build. Lastly, we will run the command shown in the following screenshot to build our container:

```
$ docker build -t my-golang-app .
```

Working with Docker Hub

So, as you can see, it was very easy to build our app, where almost all of our development time would be spent on the actual application. This will increase our productivity and bring the applications to production a lot faster. In this day and age, where agility is everything, you would have to be mad to not see the benefit.

The second type of image on Docker Hub is developed and open sourced by developers and is maintained by them individually. The easiest way to tell whether an image is official or has been developed by an individual is through the image's name. In our last example, we looked at the golang image. The name of that image is `golang`. Now, let's look at a container that I have open sourced. For this example, we will look at my `consul` image. If you want to use my image, you would call it `scottyc/consul`. As you can see, the name is different, as it calls the author name `scottyc` and then the image name, `consul`. Now, you can see the difference in the naming convention between an official image and an authored image.

Now that we have covered the different images hosted at Docker Hub, we can move on to how images get to Docker Hub. There are two different ways to get images to Docker Hub. Both ways, we need a Docker Hub account, which we will cover in the next section.

The first way is to build the image locally and simply use the `docker push` command. The second way is using automated builds, which is an awesome functionality that Docker has built into Docker Hub. We will cover this later in much more detail. At a high level, it is a **CD** (**continuous delivery**) process to build the image based on a Dockerfile that is stored in a GitHub or Bitbucket public repository.

Creating a Docker Hub account

In this topic, we will create a Docker Hub account and look at how to log in to the Docker daemon manually (we will cover how to do this with Puppet in the next chapter). So, let's start. First, we will need to go to Docker Hub (`https://hub.docker.com/`) and fill out the form on the right-hand side of the page. Just replace **yourusername** with your desired username, **you@youremail.com**, with your e-mail ID, and, of course, enter a secure password:

After that, go to your e-mail ID and confirm your account. This will then redirect you to the Docker login page. Log in and you should see the following web page:

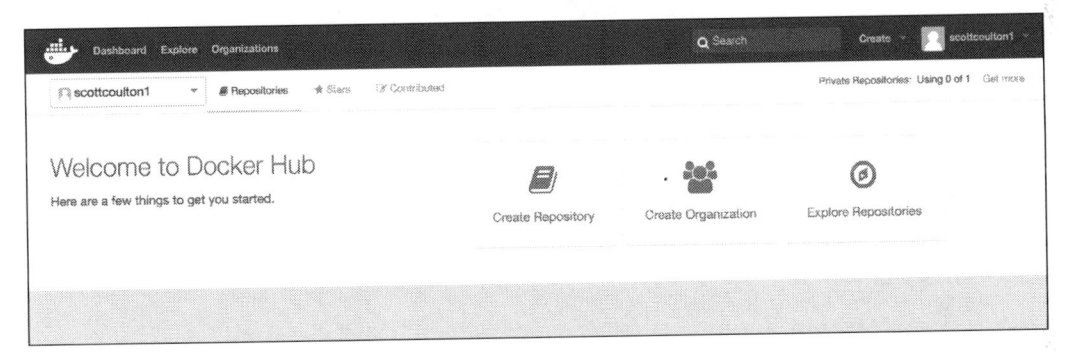

Now that we have an account, let's log in to our daemon. So, let's use `vagrant ssh` back into our Docker vagrant box. We will change to root (`sudo -i`) and then enter the `docker login` command:

Enter the username that we just created:

```
[root@node-01 ~]# docker login
Username: scottyc
Password:
```

Then, enter your password:

```
[root@node-01 ~]# docker login
Username: scottyc
Password:
Email:
```

After this, enter your e-mail ID. Once this is done, you should see the following output:

```
[root@node-01 ~]# docker login
Username: scottyc
Password:
Email: scott.coulton@gmail.com
WARNING: login credentials saved in /root/.docker/config.json
Login Succeeded
[root@node-01 ~]#
```

You have now successfully logged in the Docker daemon.

Exploring official images

In this topic, we are going to provide a quick overview of how to search for images on Docker Hub. There are two ways to do this:

- Through the Docker Hub website
- Through the command line

Let's look at the website first. If you remember, in our golang example we already used the web interface to search for an image. Let's look at another example. In this example, we will look for bitbucket, Atlassian's git server. So, we will go back to Docker Hub (https://hub.docker.com/) and enter `bitbucket` in the search field:

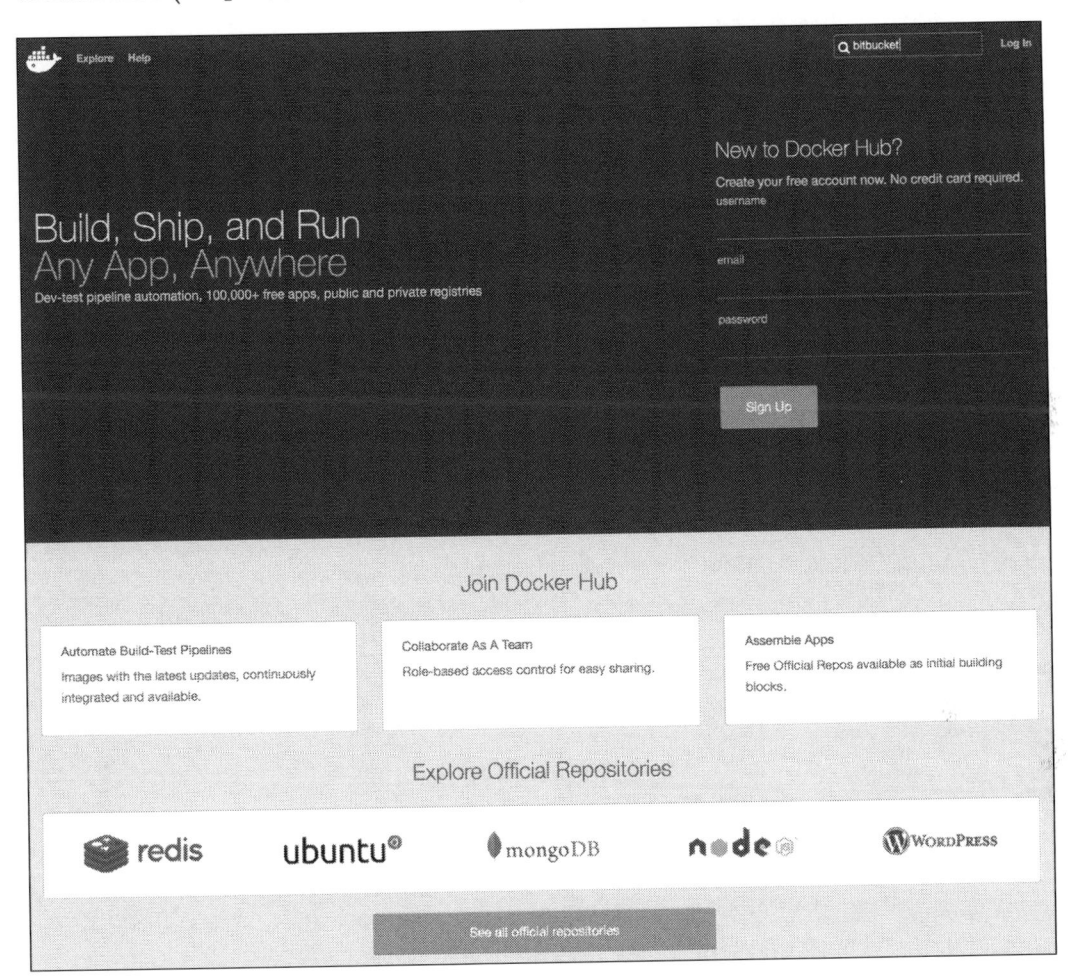

Working with Docker Hub

Our search will return the following output:

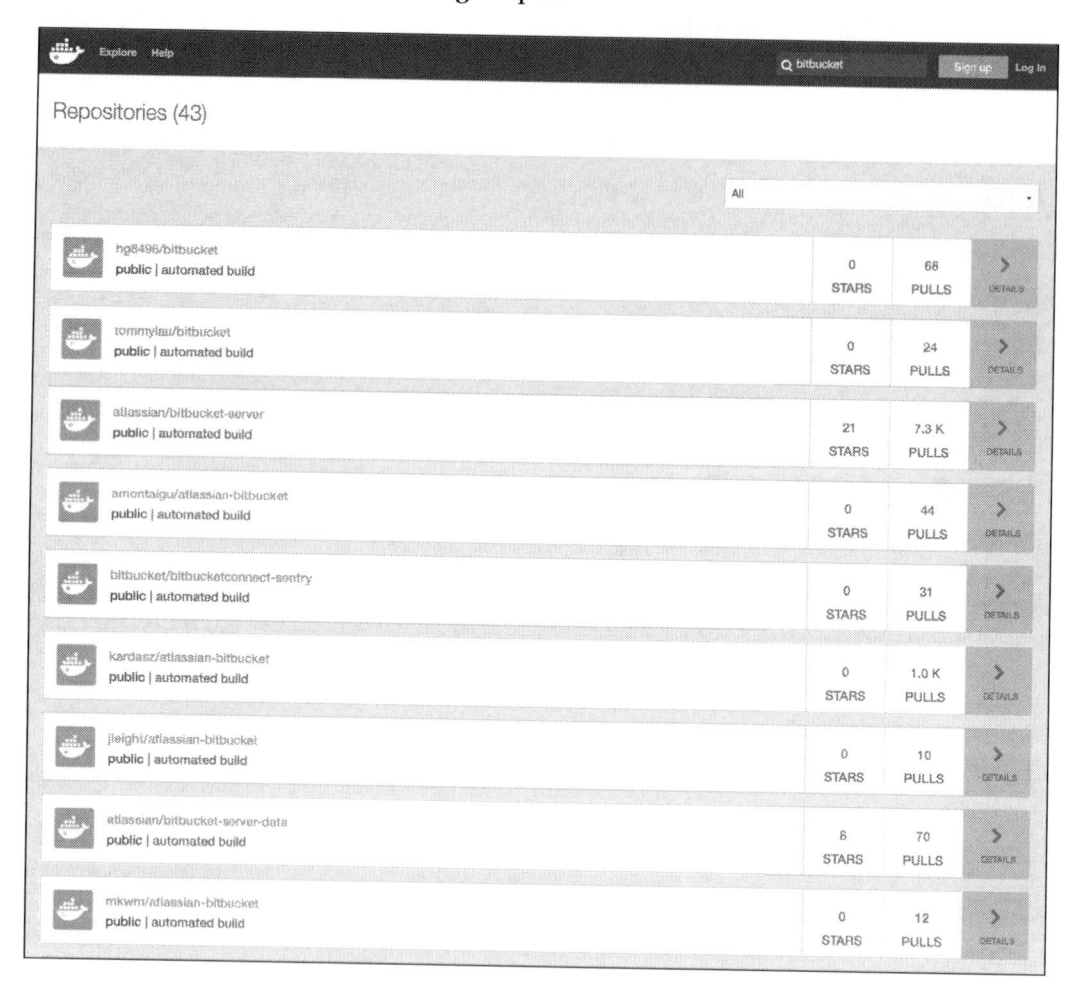

hg8496/bitbucket public | automated build	0 STARS	68 PULLS	DETAILS
tommylau/bitbucket public | automated build	0 STARS	24 PULLS	DETAILS
atlassian/bitbucket-server public | automated build	21 STARS	7.3 K PULLS	DETAILS
amontaigu/atlassian-bitbucket public | automated build	0 STARS	44 PULLS	DETAILS
bitbucket/bitbucketconnect-sentry public | automated build	0 STARS	31 PULLS	DETAILS
kardasz/atlassian-bitbucket public | automated build	0 STARS	1.0 K PULLS	DETAILS
jleight/atlassian-bitbucket public | automated build	0 STARS	10 PULLS	DETAILS
atlassian/bitbucket-server-data public | automated build	8 STARS	70 PULLS	DETAILS
mkwm/atlassian-bitbucket public | automated build	0 STARS	12 PULLS	DETAILS

As you can see from the preceding screenshot, we got 43 results. So what should we look for to choose the right image? We always look for three things, which are as follows:

- We check the number of pulls. The more people using an image, the more likely it will run with no issues.
- We also check Docker's official rating system: how many stars a repository has. Stars are awarded by other members of the community when they like the image, which is very similar to the star system on GitHub.

Chapter 2

- We check whether the repo has a Dockerfile. This gives you peace of mind about how the image is built. You can see all the commands that were run to complete the build.

Using the three metrics, let's pick an image. Looking at the results, **atlassian/bitbucket-server** looks good, with 21 stars and 7.3k pulls. So, let's click on the repo and look for a Dockerfile:

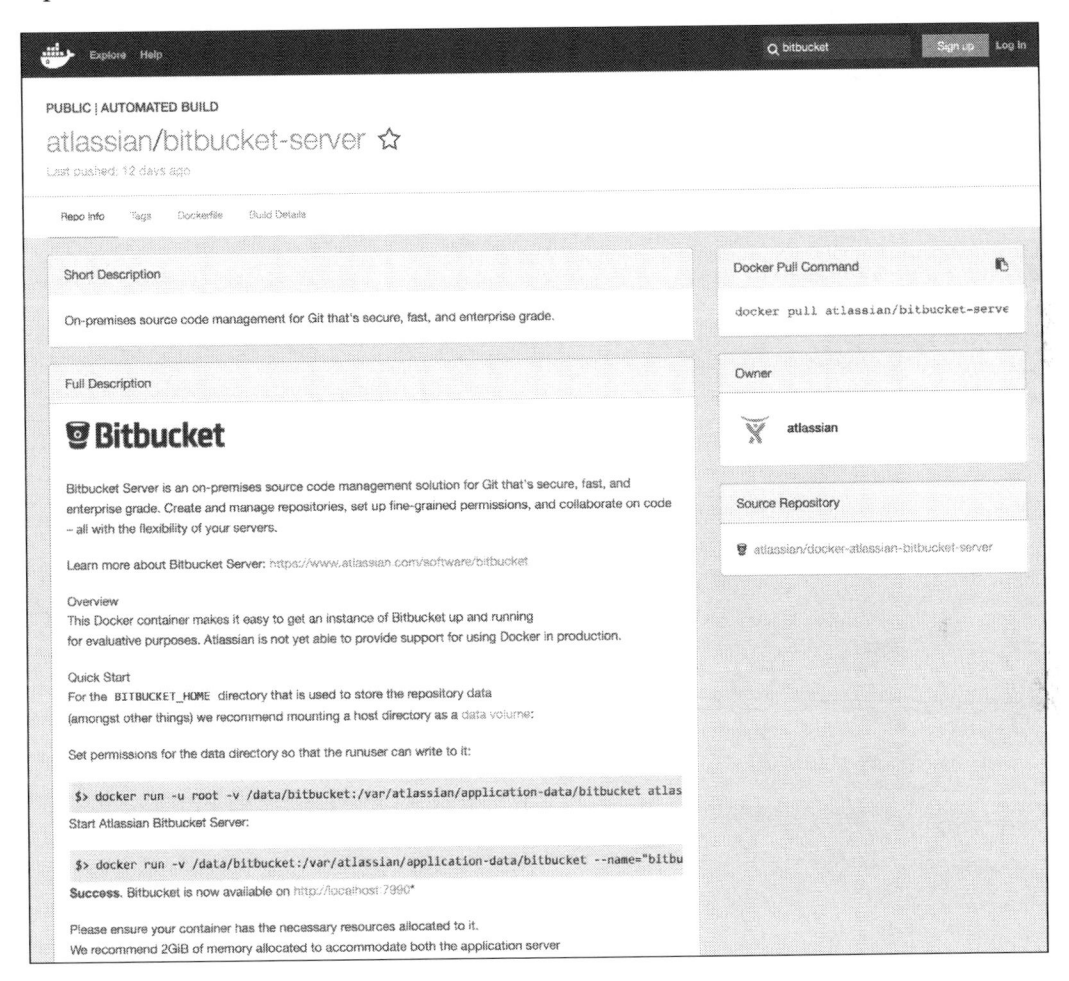

![Screenshot of the atlassian/bitbucket-server page on Docker Hub showing the Short Description, Full Description with Bitbucket overview and Quick Start commands, Docker Pull Command, Owner (atlassian), and Source Repository (atlassian/docker-atlassian-bitbucket-server).]

If we click on the **Dockerfile** tab under the main image title, it takes us to the Dockerfile page. Not every repository has a Dockerfile; however, this does not mean that it's a bad image. This just means that it will take more testing before you would be able use it in production. Some authors, such as *Jess (Jessie Frazelle)* from Docker, have their Dockerfiles on their GitHub page. She has awesome images on Docker Hub and the Dockerfiles can be found at `https://github.com/jfrazelle/dockerfiles`. Alright, back to our example. As you can see in the following screenshot, there is a Dockerfile:

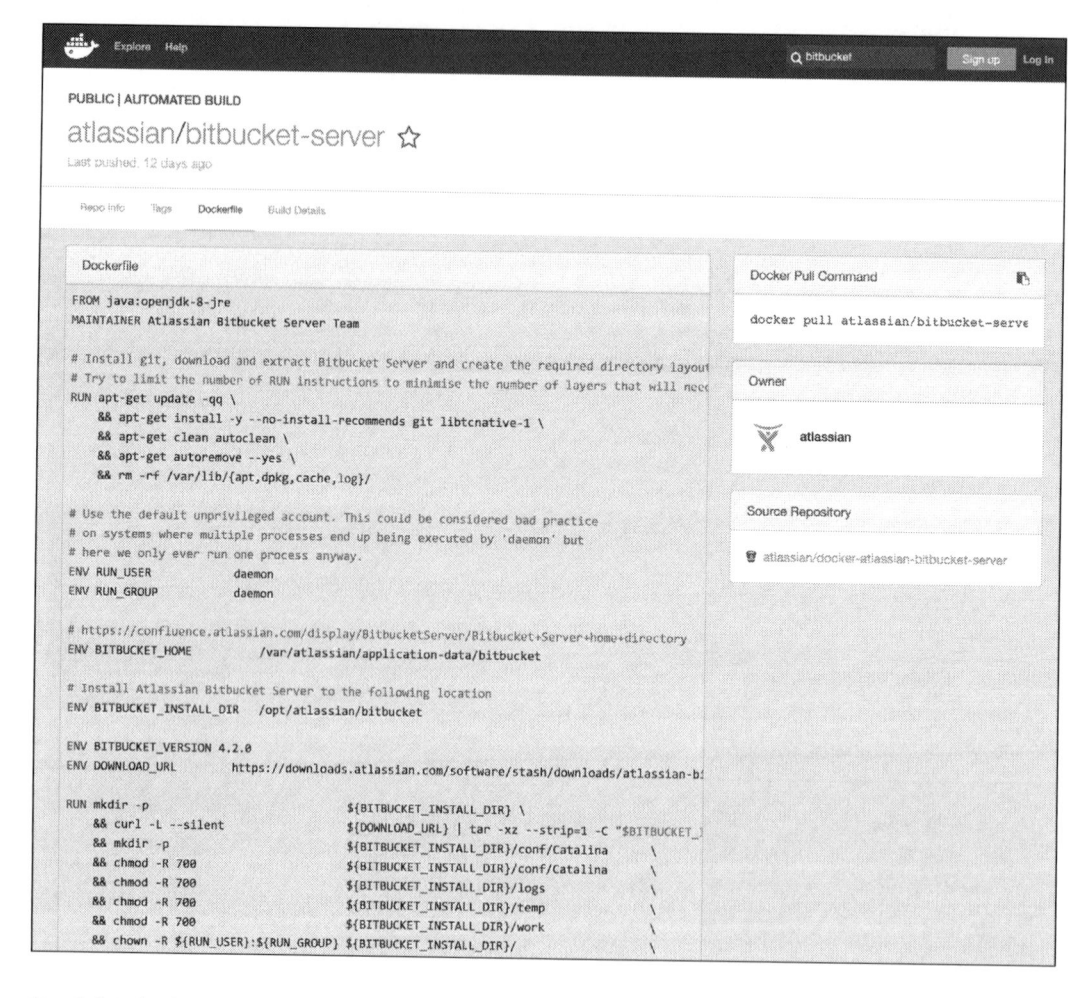

So, I think this is the winner!!!!

Now, let's do the same search from the command line. In the command line, type `docker search bitbucket`, and the search will return the following output:

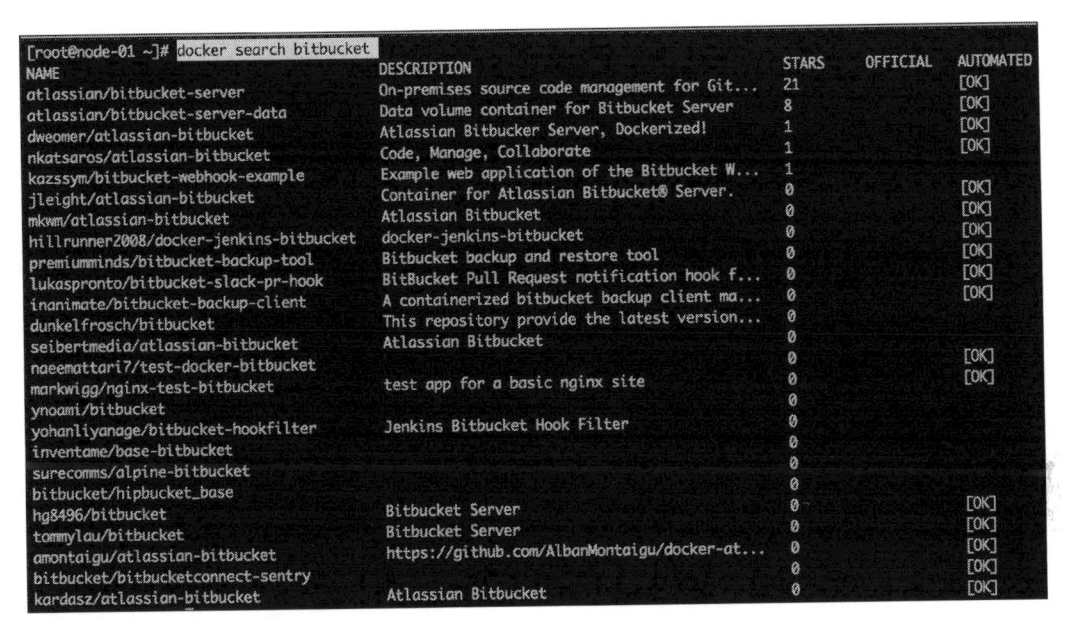

```
[root@node-01 ~]# docker search bitbucket
NAME                                          DESCRIPTION                                     STARS     OFFICIAL   AUTOMATED
atlassian/bitbucket-server                    On-premises source code management for Git...   21                   [OK]
atlassian/bitbucket-server-data               Data volume container for Bitbucket Server      8                    [OK]
dweomer/atlassian-bitbucket                   Atlassian Bitbucker Server, Dockerized!         1                    [OK]
nkatsaros/atlassian-bitbucket                 Code, Manage, Collaborate                       1                    [OK]
kazssym/bitbucket-webhook-example             Example web application of the Bitbucket W...   1
jleight/atlassian-bitbucket                   Container for Atlassian Bitbucket® Server.      0                    [OK]
mkwm/atlassian-bitbucket                      Atlassian Bitbucket                             0                    [OK]
hillrunner2008/docker-jenkins-bitbucket       docker-jenkins-bitbucket                        0                    [OK]
premiuminds/bitbucket-backup-tool             Bitbucket backup and restore tool               0                    [OK]
lukaspronto/bitbucket-slack-pr-hook           BitBucket Pull Request notification hook f...   0                    [OK]
inanimate/bitbucket-backup-client             A containerized bitbucket backup client ma...   0                    [OK]
dunkelfrosch/bitbucket                        This repository provide the latest version...   0
seibertmedia/atlassian-bitbucket              Atlassian Bitbucket                             0
naeemattari7/test-docker-bitbucket                                                            0                    [OK]
markwigg/nginx-test-bitbucket                 test app for a basic nginx site                 0                    [OK]
ynoami/bitbucket                                                                              0
yohanliyanage/bitbucket-hookfilter            Jenkins Bitbucket Hook Filter                   0
inventame/base-bitbucket                                                                      0
surecomms/alpine-bitbucket                                                                    0
bitbucket/hipbucket_base                                                                      0
hg8496/bitbucket                              Bitbucket Server                                0                    [OK]
tommylau/bitbucket                            Bitbucket Server                                0                    [OK]
amontaigu/atlassian-bitbucket                 https://github.com/AlbanMontaigu/docker-at...   0                    [OK]
bitbucket/bitbucketconnect-sentry                                                             0                    [OK]
kardasz/atlassian-bitbucket                   Atlassian Bitbucket                             0                    [OK]
```

As you can see, it has returned the same information, and the only thing missing is the number of pulls. Again, it looks like we will use **atlassian/bitbucket-server**.

Automated builds in Docker Hub

In this topic, we are going to look at how automated builds work at a high level, and how to publish an image via the push method on Docker Hub.

Working with Docker Hub

Automated builds

In Docker Hub, we have two ways to publish images: via a simple push method or via an automated build. In this topic, we will cover automated builds. First, we will look at the flow of an automated build. In this example, we will be using GitHub, but you can also use Bitbucket. So, the first thing that we need to do is link our Docker Hub account to our GitHub account. This is done by navigating to **Settings | Linked Accounts & Services**:

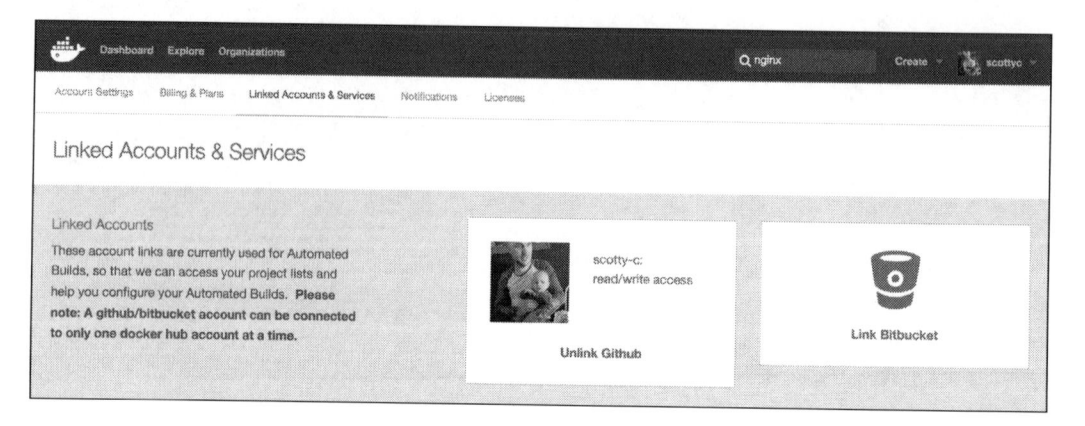

Just follow the prompts to link the accounts.

Once this is completed, let's go to our GitHub account and create a repo. I am going to use the one that I have already set up:

RVM

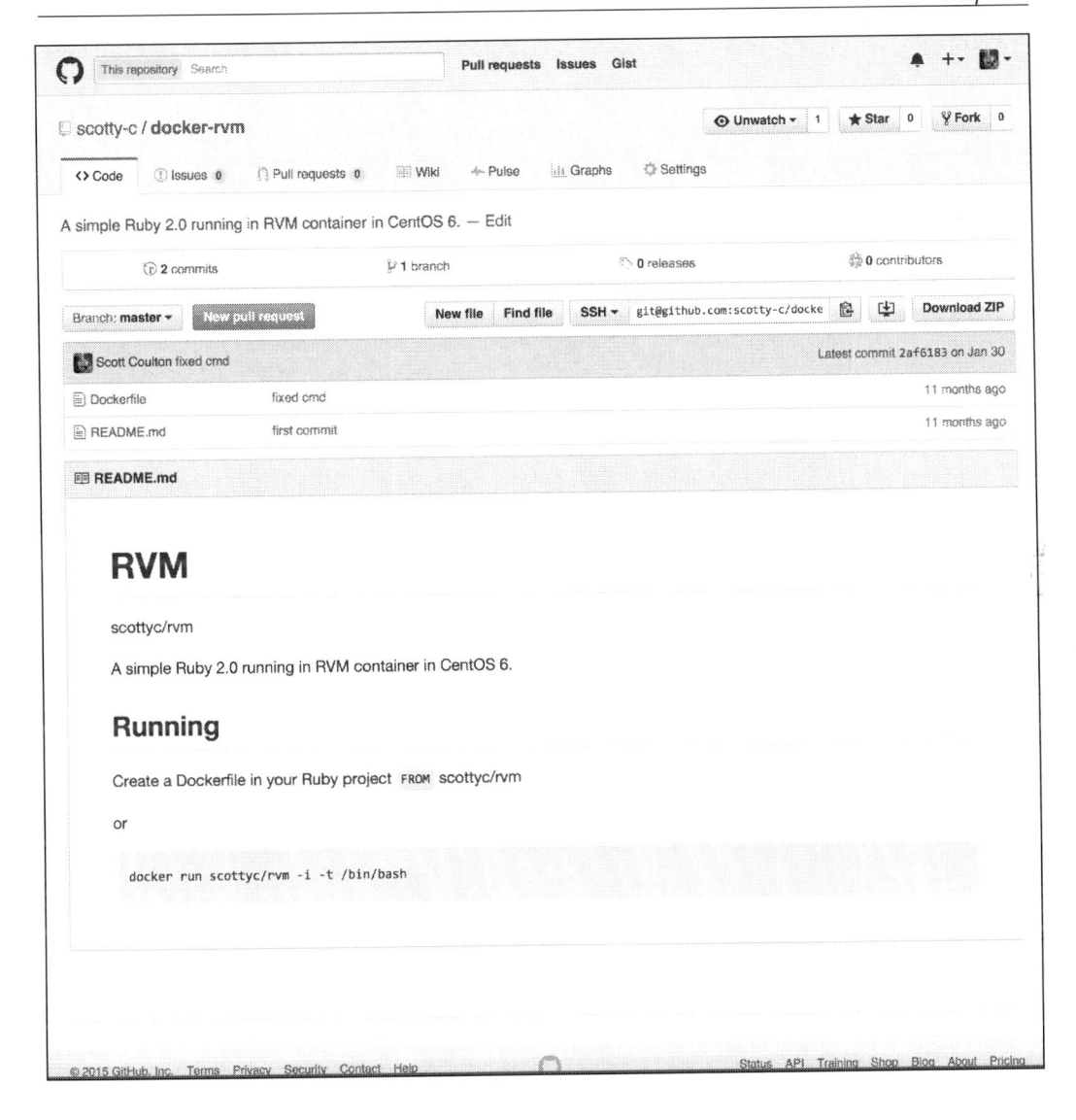

scottyc/rvm

A simple Ruby 2.0 running in RVM container in CentOS 6.

Running

Create a Dockerfile in your Ruby project `FROM` scottyc/rvm

or

```
docker run scottyc/rvm -i -t /bin/bash
```

Working with Docker Hub

As you can see in the preceding screenshot, the repo contains a Dockerfile. Now, let's looks at the same repo except Docker Hub:

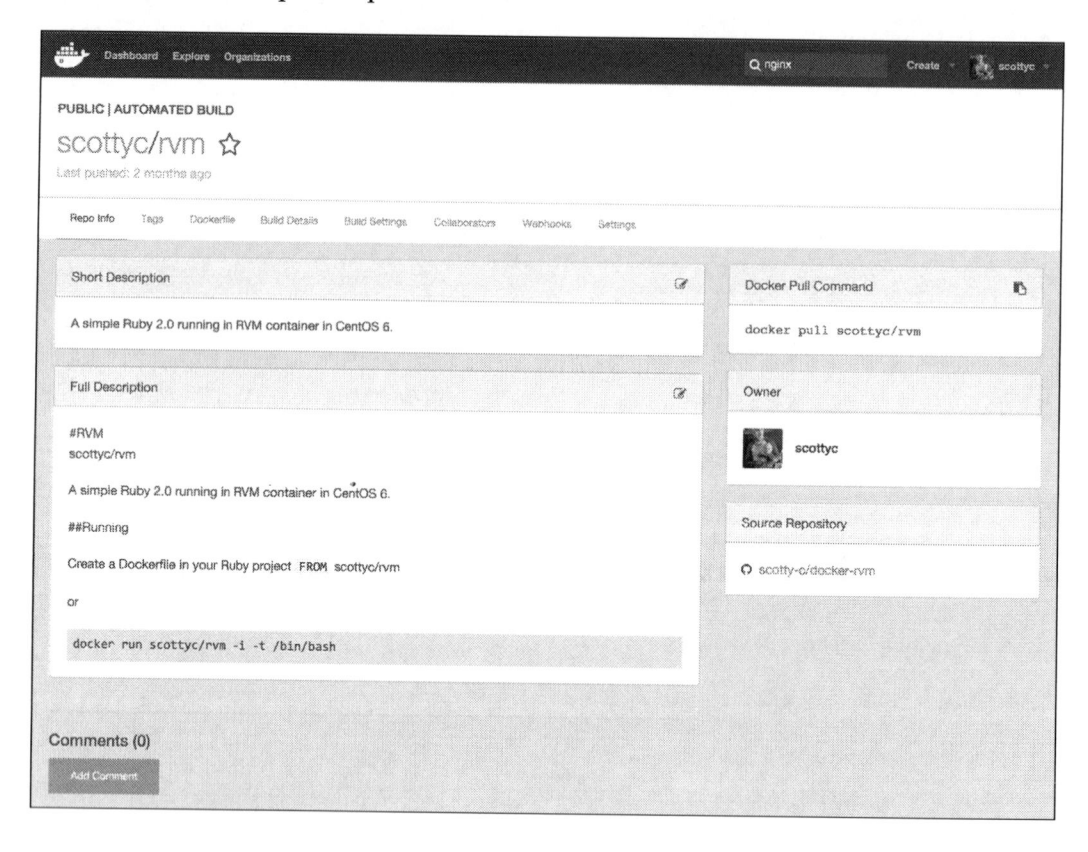

After this, we will look at the **Build Details** tab:

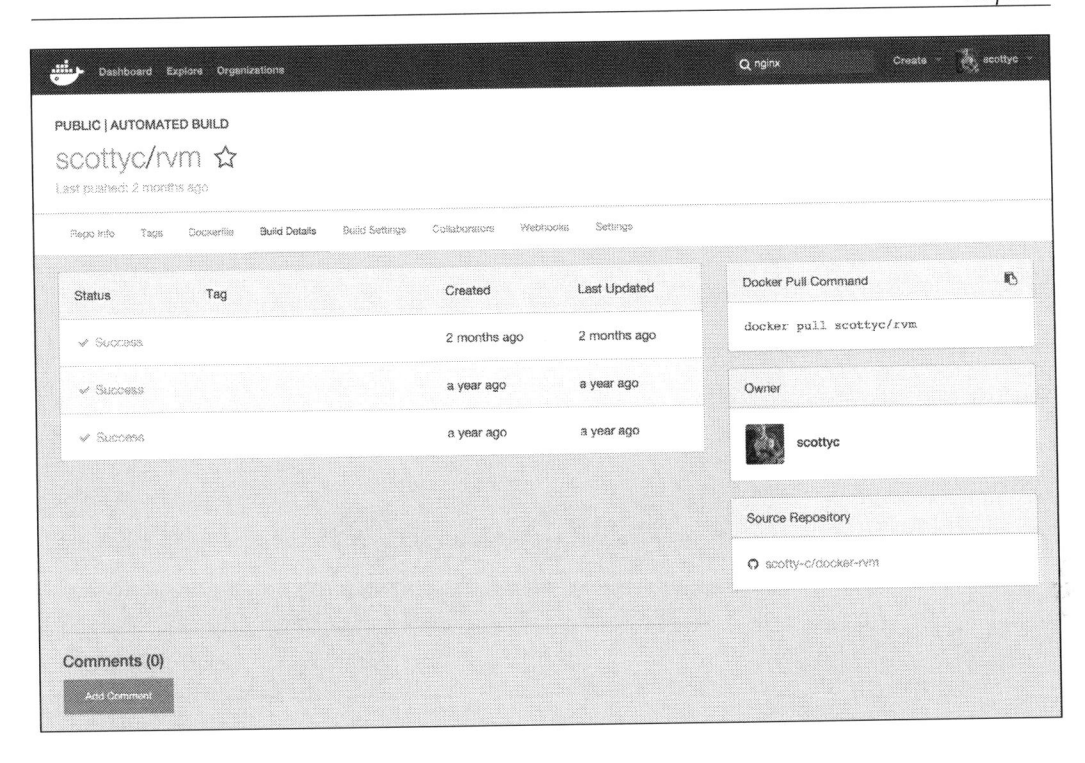

So, how does that build automate ? Well, it is quite simple. Every time we check in a change to the GitHub repo, it will trigger web hooks at Docker Hub. When Docker Hub receives the trigger, it will grab the Dockerfile and build the image. Docker Hub will take care of things such as version numbers for us with every build. So, at a high level, this is how automated builds work.

Pushing to Docker Hub

This is a quite simple way to get an image to Docker Hub, but the downside is that there is no automated build process and the Dockerfile does not get placed in the Docker Hub repo automatically. So, in this example, we will assume that we have created an image called `scottyc/super_app`. To push this to Docker Hub, we simply type `docker push scottyc/super_app` in the terminal. Note that the Docker daemon needs to be logged in at the time of the push.

Working with official images

Now that we know how Docker Hub serves images to us, let's look at how to integrate them into our code via three methods. The first will be a Dockerfile, the second will be in the `docker-compose.yaml` file, and the last will be straight into a Puppet manifest.

Dockerfiles

In this topic, we will look at using nginx in a basic Dockerfile. In a Dockerfile, we need to add a few things. The first is the image that we are basing our application on; for us it will be nginx. The second is a maintainer. It should look like as shown in the following screenshot:

```
1  FROM nginx
2
3  MAINTAINER Scott Coulton
```

As the base nginx image has already got port 80 and 443 exposed, we will not need that configuration for our Dockerfile. The next thing we will add is a simple `run` command to update the packages in the container. As its base OS is Debian, we will add the command shown on line **5** in the following screenshot:

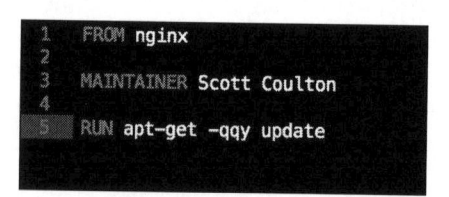

```
1  FROM nginx
2
3  MAINTAINER Scott Coulton
4
5  RUN apt-get -qqy update
```

As we are building a simple application, this is all that we are going to add to our Dockerfile. There are heaps of configurations that can be done with a Dockerfile.

> If you would like to read about Dockerfiles, you can do so at https://docs.docker.com/engine/reference/builder/.

Now, let's build our image. You will note that `server.yaml` in our Vagrant repo already has port `80` forwarding to port `8080`, so we won't need to make any changes there. Copy the Dockerfile that we created into the root of your Vagrant repo. Then, let's start our vagrant box with vagrant up from our terminal. Then, use `vagrant ssh` once the box is up. Let's change to root (`sudo -i`). Then, if we change directories to `/vagrant`, we should see our Dockerfile. Now, let's build our image with the command, `docker build -t YOUR AUTHOR NAME/nginx` . (note that . is part of the command). You will get the following output on your terminal:

```
[root@node-01 vagrant]# docker build -t scottyc/nginx .
Sending build context to Docker daemon 98.82 kB
Step 1 : FROM nginx
latest: Pulling from library/nginx

9ee13ca3b908: Pull complete
23cb15b0fcec: Pull complete
62df5e17dafa: Pull complete
d65968c1aa44: Pull complete
f5bb1dddc876: Pull complete
1526247f349d: Pull complete
2e518e3d3fad: Pull complete
0e07123e6531: Pull complete
21656a3c1256: Pull complete
f608475c6c65: Pull complete
1b6c0a20b353: Pull complete
5328fdfe9b8e: Pull complete
Digest: sha256:a79db4b83c0dbad9542d5442002ea294aa77014a3dfa67160d8a55874a5520cc
Status: Downloaded newer image for nginx:latest
 ---> 5328fdfe9b8e
Step 2 : MAINTAINER Scott Coulton
 ---> Running in 5c30c184f0d6
 ---> d2843a2a5a53
Removing intermediate container 5c30c184f0d6
Step 3 : RUN apt-get -qqy update
 ---> Running in b8fd1c675494
 ---> 821117a98fcd
Removing intermediate container b8fd1c675494
Successfully built 821117a98fcd
```

Next, let's test our image and spin up a container with the following command:

```
docker run -d -p 80:80 --name=nginx YOUR AUTHOR NAME/nginx
```

Working with Docker Hub

If it was successful, we should get the nginx default page in your browser at `127.0.0.1:8080`, as follows:

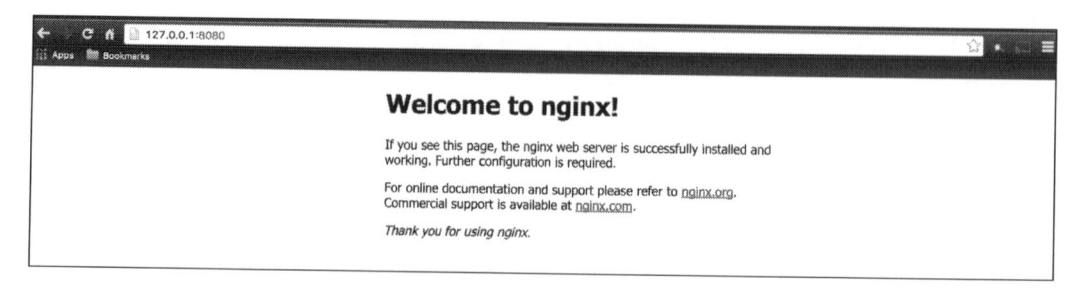

![Welcome to nginx! If you see this page, the nginx web server is successfully installed and working. Further configuration is required. For online documentation and support please refer to nginx.org. Commercial support is available at nginx.com. Thank you for using nginx.]

Docker Compose

Now, we are going to deploy the same nginx image with Docker Compose. We are going to run Docker Compose at a high level in this topic just to get an understanding of the technology. We will look at it in depth in another chapter of this book. The first thing we need to do is install Docker Compose.

> At the time of writing this book, my pull request is still open, so we will have to use my branch of Gareth's module.

To do this, let's modify our puppetfile in the Vagrant repo with the commands shown in the following screenshot:

```
#!/usr/bin/ruby env

require "socket"
$hostname = Socket.gethostname

forge 'http://forge.puppetlabs.com'


mod 'puppetlabs/stdlib', '4.1.0'
mod 'puppetlabs/apt', '2.2.1'
mod 'stahnma/epel'
mod 'garethr/docker', :git => "https://github.com/scotty-c/garethr-docker.git"
mod 'stankevich/python'
```

So, in the Puppetfile we added a new module dependency, `stankevich/python`, as Docker Compose is written in Python. We also updated our `epel` module to use the latest. Just to get a fresh working environment, we will run the command, `vagrant destroy && vagrant up`, in our terminal. Once the box is up, we will use `vagrant ssh` and then change to root (`sudo -i`). We will then change the directory to `/vagrant` and type `docker-compose`.

If the build was successful, we will see the following screen:

```
[vagrant@node-01 ~]$ sudo -i
[root@node-01 ~]# docker-compose
Define and run multi-container applications with Docker.

Usage:
  docker-compose [-f=<arg>...] [options] [COMMAND] [ARGS...]
  docker-compose -h|--help

Options:
  -f, --file FILE            Specify an alternate compose file (default: docker-compose.yml)
  -p, --project-name NAME    Specify an alternate project name (default: directory name)
  --x-networking             (EXPERIMENTAL) Use new Docker networking functionality.
                             Requires Docker 1.9 or later.
  --x-network-driver DRIVER  (EXPERIMENTAL) Specify a network driver (default: "bridge").
                             Requires Docker 1.9 or later.
  --verbose                  Show more output
  -v, --version              Print version and exit

Commands:
  build              Build or rebuild services
  help               Get help on a command
  kill               Kill containers
  logs               View output from containers
  pause              Pause services
  port               Print the public port for a port binding
  ps                 List containers
  pull               Pulls service images
  restart            Restart services
  rm                 Remove stopped containers
  run                Run a one-off command
  scale              Set number of containers for a service
  start              Start services
  stop               Stop services
  unpause            Unpause services
  up                 Create and start containers
  migrate-to-labels  Recreate containers to add labels
  version            Show the Docker-Compose version information
```

Now, let's create `docker-compose.yaml`:

```
1  nginx:
2      image: nginx
3      hostname: nginx
4      ports:
5          - "80:80"
```

Working with Docker Hub

As you can see, we used the official image, gave the container a name `nginx`, and exposed ports **80:80** again to be able to hit the nginx page. So, if we copy our `docker-compose.yml` file to the root of the Vagrant directory, log in to our vagrant box, and change the directory to root (`vagrant ssh`, then `sudo -i`), we will be able to change the directory to `/vagrant` again. Now, run `docker-compose up -d`. We will get the following output after running it:

```
[root@node-01 vagrant]# docker-compose up -d
Pulling nginx (nginx:latest)...
latest: Pulling from library/nginx
9ee13ca3b908: Pull complete
23cb15b0fcec: Pull complete
62df5e17dafa: Pull complete
d65968c1aa44: Pull complete
f5bb1dddc876: Pull complete
1526247f349d: Pull complete
2e518e3d3fad: Pull complete
0e07123e6531: Pull complete
21656a3c1256: Pull complete
f608475c6c65: Pull complete
1b6c0a20b353: Pull complete
5328fdfe9b8e: Pull complete
Digest: sha256:a79db4b83c0dbad9542d5442002ea294aa77014a3dfa67160d8a55874a5520cc
Status: Downloaded newer image for nginx:latest
Creating vagrant_nginx_1
```

We can then go to our web browser and visit the nginx page at `127.0.0.1:8080`:

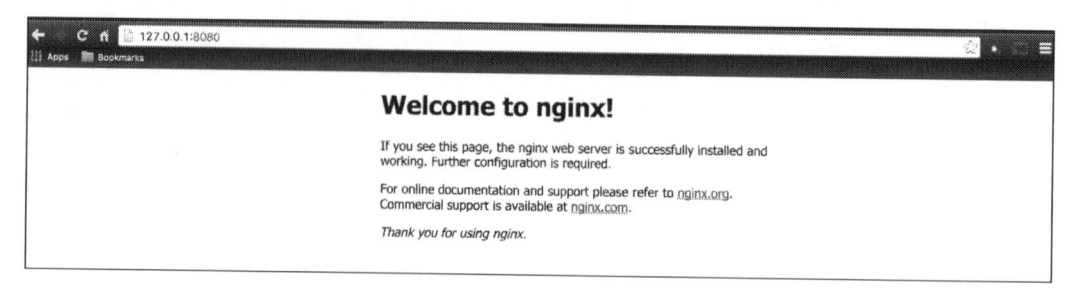

Welcome to nginx!

If you see this page, the nginx web server is successfully installed and working. Further configuration is required.

For online documentation and support please refer to nginx.org.
Commercial support is available at nginx.com.

Thank you for using nginx.

> If you want to read more about Docker Compose, go to `https://docs.docker.com/compose/`.

Puppet manifest

In this section, we are going to build the same ngnix container with a simple Puppet manifest. This is just a proof of concept. In the next chapter, we will write a full module. This is just to give us a foundation and understanding of how Puppet interacts with Docker. So, in our Vagrant repo, let's modify `manifest/default.pp`. The file should contain the following code:

```
node 'node-01' {

  include docker

  docker::run { 'nginx':
    image    => 'nginx',
    ports    => ['80', '80'],
    hostname => 'nginx',
  }
}
```

We will then open our terminal at the root of our Vagrant repo and run `vagrant provision`. Note that you should have no other containers running at this time. You will see the following output, which shows that Puppet provisioned a Docker container called nginx:

```
==> node-01: Running provisioner: puppet...
==> node-01: Running Puppet with default.pp...
==> node-01: Info: Loading facts
==> node-01: Info: Loading facts
==> node-01: Info: Loading facts
==> node-01: Notice: Compiled catalog for localhost in environment production in 1.28 seconds
==> node-01: Info: Applying configuration version '1451523602'
==> node-01: Notice: /Stage[main]/Main/Node[node-01]/Docker::Run[nginx]/File[/etc/systemd/system/docker-nginx.service]/ensure: created
==> node-01: Info: /Stage[main]/Main/Node[node-01]/Docker::Run[nginx]/File[/etc/systemd/system/docker-nginx.service]: Scheduling refresh of Exec[docker-systemd-reload]
==> node-01: Info: /Stage[main]/Main/Node[node-01]/Docker::Run[nginx]/File[/etc/systemd/system/docker-nginx.service]: Scheduling refresh of Service[docker-nginx]
==> node-01: Notice: /Stage[main]/Docker::Systemd_reload/Exec[docker-systemd-reload]: Triggered 'refresh' from 1 events
==> node-01: Notice: /Stage[main]/Main/Node[node-01]/Docker::Run[nginx]/Service[docker-nginx]/ensure: ensure changed 'stopped' to 'running'
==> node-01: Info: /Stage[main]/Main/Node[node-01]/Docker::Run[nginx]/Service[docker-nginx]: Unscheduling refresh on Service[docker-nginx]
==> node-01: Notice: Finished catalog run in 1.14 seconds
```

We can then check our browser again at `127.0.0.1:8080`. We will get the nginx page again:

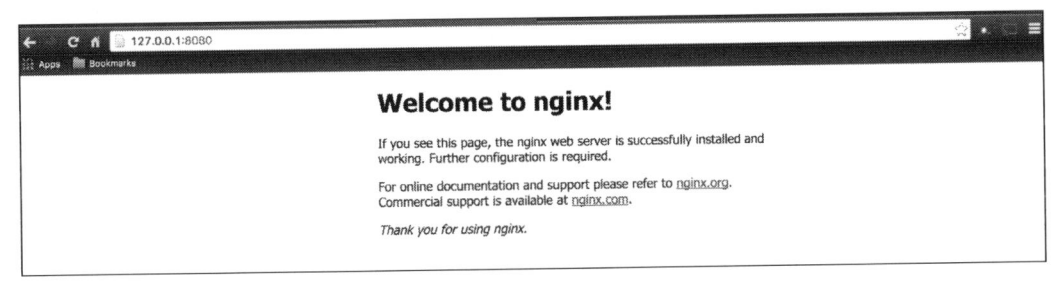

Welcome to nginx!

If you see this page, the nginx web server is successfully installed and working. Further configuration is required.

For online documentation and support please refer to nginx.org.
Commercial support is available at nginx.com.

Thank you for using nginx.

Summary

In this chapter, we covered a lot about the Docker Hub ecosystem. We discussed what official images are, how automated builds work, and of course, how to work with images in three different ways. After working through this chapter, we now have the tools in our tool belt to build our first application with Puppet.

In the next chapter, we will write our first Puppet module to create a Docker container and we will look at writing rspec-puppet unit tests to make sure that our module does what it's meant to do.

3
Building a Single Container Application

In this chapter, we are going to write our first module to deploy our first containerized application. The application that we are going to deploy is Consul from HashiCorp (`https://www.consul.io/`). We will talk about Consul a little later in the chapter. The first thing we will look at is how to construct a Puppet module with the correct file structure, unit tests, and gems. Once we have our module skeleton, we will look at the two ways to deploy Consul with Puppet in a container. The first will be to use resource declarations in a manifest and the second will be to use Docker Compose as a template `.erb` file. These are the topics that we will cover in this chapter:

- Building a Puppet module skeleton
- Coding using resource declarations
- Coding using `.erb` files

Building a Puppet module skeleton

One of the most important things in development is having a solid foundation. Writing a Puppet module is no different. This topic is extremely important for the rest of the book, as we will be reusing the code over and over again to build all our modules from now on. We will first look at how to build a module with the Puppet module generator. Once we have our module skeleton, we will look at its construct. We will look at the plumbing Puppet uses with Ruby, and lastly, at basic unit tests.

The Puppet module generator

One of the best things about working with Puppet is the number of tools out there, both from the community and from puppetlabs itself. The Puppet module generator is a tool that is developed by puppetlabs and follows the best practices to create a module skeleton. The best thing about this tool is that it is bundled with every Puppet agent install. So, we don't need to install any extra software. Let's log in to our Vagrant box that we built in the last chapter. Let's change directory to the root of our Vagrant repo and then use the `vagrant up && vagrant ssh` command. Now that we are logged in to the box, let's sudo to root (`sudo -i`) and change the directory to `/vagrant`. The reason for this is that this folder will be mapped to our local box. Then, we can use our favorite text editor later in the chapter. Once we're in `/vagrant`, we can run the command to build our Puppet module skeleton. The `puppet module generate <AUTHOR>-consul` command for me will look like this: `puppet module generate scottyc-consul`.

The script will then ask a few questions such as the version, author name, description, where the source lives, and so on. These are very important questions that are to be considered when you want to publish a module to the Puppet Forge (`https://forge.puppetlabs.com/`), but for now, let's just answer the questions as per the following example:

```
[root@node-01 ~]# puppet module generate scottyc-consul
We need to create a metadata.json file for this module.  Please answer the
following questions; if the question is not applicable to this module, feel free
to leave it blank.

Puppet uses Semantic Versioning (semver.org) to version modules.
What version is this module?  [0.1.0]
-->

Who wrote this module?  [scottyc]
-->

What license does this module code fall under?  [Apache 2.0]
-->

How would you describe this module in a single sentence?
--> This is a module the runs Conusl in a Docker container

Where is this module's source code repository?
-->

Where can others go to learn more about this module?
-->

Where can others go to file issues about this module?
-->

----------------------------------------
{
  "name": "scottyc-consul",
  "version": "0.1.0",
  "author": "scottyc",
  "summary": "This is a module the runs Conusl in a Docker container",
  "license": "Apache 2.0",
  "source": "",
  "project_page": null,
  "issues_url": null,
  "dependencies": [
    {"name":"puppetlabs-stdlib","version_requirement":">= 1.0.0"}
  ]
}
----------------------------------------

About to generate this metadata; continue? [n/Y]
--> y

Notice: Generating module at /root/scottyc-consul...
Notice: Populating templates...
Finished; module generated in scottyc-consul.
scottyc-consul/Gemfile
scottyc-consul/Rakefile
scottyc-consul/manifests
scottyc-consul/manifests/init.pp
scottyc-consul/spec
scottyc-consul/spec/classes
scottyc-consul/spec/classes/init_spec.rb
scottyc-consul/spec/spec_helper.rb
scottyc-consul/tests
scottyc-consul/tests/init.pp
scottyc-consul/README.md
scottyc-consul/metadata.json
[root@node-01 ~]#
```

Building a Single Container Application

Now that we have our Puppet module skeleton, we should look at what the structure looks like:

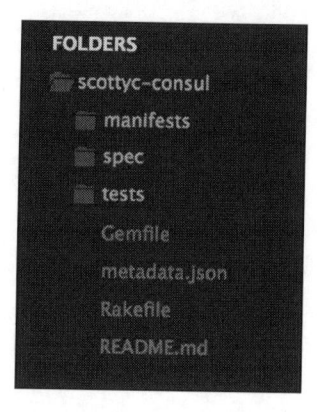

```
FOLDERS
  scottyc-consul
    manifests
    spec
    tests
    Gemfile
    metadata.json
    Rakefile
    README.md
```

Now, we are going to add a few files to help us with unit tests. The first file is `.fixtures.yml`. This file is used by `spec-puppet` to pull down any module dependencies into the `spec/fixtures` directory when we run our unit tests. For this module, the `.fixtures.yml` file should look like the one shown in the following screenshot:

```yaml
fixtures:
    symlinks:
        consul: "#{source_dir}"
    forge_modules:
        vcsrepo: puppetlabs/vcsrepo
        stdlib: puppetlabs/stdlib
        golang: scottyc/golang
        python: stankevich/python
    repositories:
        docker: https://github.com/scotty-c/garethr-docker.git
```

The next file that we are going to add is a `.rspec` file. This is the file that `rspec-puppet` uses when it requires `spec_helper`, and it sets the pattern for our unit test folder structure. The file contents should look as shown in this screenshot:

```
--require spec_helper
--pattern spec/*/*_spec.rb
```

Now that we have our folder structure, let's install the gems that we need to run our unit tests. My personal preference is to install the gems on the vagrant box; if you want to use your local machine, that's fine as well. So, let's log in to our vagrant box (cd into the root of our Vagrant repo, use the `vagrant ssh` command, and then change the directory to root using `sudo -i`). First, we will install Ruby with `yum install -y ruby`. Once that is complete, let's cd into `/vagrant/<your modules folder>` and then run `gem install bundler && bundle install`. You should get the following output:

```
[root@node-01 scottyc-consul]# bundle install
Don't run Bundler as root. Bundler can ask for sudo if it is needed, and installing your bundle as root will break this application for all non-root users on this machine.
Fetching gem metadata from https://rubygems.org/..........
Fetching version metadata from https://rubygems.org/..
Resolving dependencies...
Rubygems 2.0.14 is not threadsafe, so your gems will be installed one at a time. Upgrade to Rubygems 2.1.0 or higher to enable parallel gem installation.
Installing rake 10.5.0
Installing CFPropertyList 2.2.8
Installing diff-lcs 1.2.5
Installing facter 2.4.4
Installing json_pure 1.8.3
Installing metaclass 0.0.4
Installing puppet-lint 1.1.0
Installing rspec-support 3.4.1
Using bundler 1.11.2
Installing puppet-syntax 2.0.0
Installing hiera 3.0.5
Installing mocha 1.1.0
Installing rspec-core 3.4.1
Installing rspec-expectations 3.4.0
Installing rspec-mocks 3.4.1
Installing puppet 4.3.1
Installing rspec 3.4.0
Installing rspec-puppet 2.3.0
Installing puppetlabs_spec_helper 1.0.1
Bundle complete! 4 Gemfile dependencies, 19 gems now installed.
Use `bundle show [gemname]` to see where a bundled gem is installed.
```

As you can see from the preceding screenshot, we got some warnings. This is because we ran `gem install` as the root. We would not do that on a production system, but as this is our development box, it won't pose an issue. Now that we have all the gems that we need for our unit tests, let's add some basic facts to `/spec/classes/init_spec.rb`. The facts we are going to add are `osfamily` and `operatingsystemrelease`. So, the file will look as shown in this screenshot:

```
1  require 'spec_helper'
2  describe 'consul' do
3
4    let(:facts) { {:osfamily => 'RedHat', :operatingsystemrelease => 'RedHat Linux release 7.0'}}
5
6    context 'with defaults for all parameters' do
7      it { should contain_class('consul') }
8    end
9  end
10
```

Building a Single Container Application

The last file that we will edit is the `metadata.json` file in the root of the repo. This file defines our module dependencies. For this module, we have one dependency, `docker`, so we need to add that at the bottom of the `metadata.json` file, as shown in the following screenshot:

```
{
  "name": "scottyc-consul",
  "version": "0.1.0",
  "author": "scottyc",
  "summary": "This is a module the runs Conusl in a Docker container",
  "license": "Apache 2.0",
  "source": "",
  "project_page": null,
  "issues_url": null,
  "dependencies": [
    { "name": "garethr/docker", "version_requirement": ">= 4.0.0" }
  ]
}
```

The last thing we need to do is put everything in its place inside our Vagrant repo. We do that by creating a folder called `modules` in the root of our Vagrant repo. Then, we issue the `mv <AUTHOR>-consul/ modules/consul` command. Note that we removed the author name because we need the module to replicate what it would look like on a Puppet master. Now that we have our basic module skeleton ready, we can start with some coding.

Coding using resource declarations

In this section, we are going to use our module skeleton to build our first Docker application. We are going to write it using standard Puppet manifests.

But first, why is the first module that we are writing Consul? I chose this application for a few reasons. First, Consul has a lot of awesome features such as service discovery and health checks, and can be used as a key/value store. The second reason is that we will use all the features I just mentioned later in the book. So, it will come in handy when we look at Docker Swarm.

File structures

Let's create two new files, `install.pp` and `params.pp`, in the `manifests` folder. The structure should look as shown in the following screenshot:

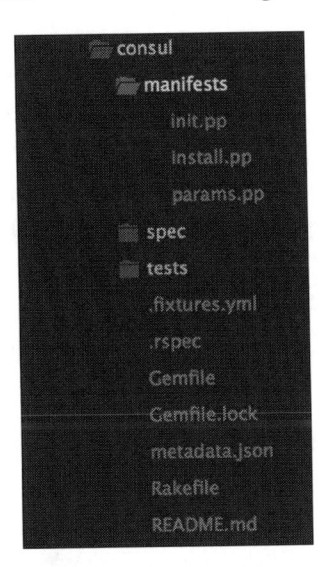

```
consul
  manifests
    init.pp
    install.pp
    params.pp
  spec
  tests
  .fixtures.yml
  .rspec
  Gemfile
  Gemfile.lock
  metadata.json
  Rakefile
  README.md
```

Writing our module

Let's start writing our module. We will start with `init.pp`; this module is not going to be very complex as we are only going to add a few lines of code and some parameters. As you can see in the preceding screenshot, we created three files in the `manifests` directory. When I write a module, I always like to start at `params.pp`, as it gives me a good starting structure to work for the code that provides the module logic. So, let's look at `params.pp` for this module, which is shown in the following screenshot:

```
class consul::params {

    $docker_version         = '1.9.1-1.el7.centos'
    $docker_tcp_bind        = 'tcp://127.0.0.1:4243'
    $docker_image           = 'scottyc/consul'
    $container_hostname     = 'consul'
    $consul_advertise       = $::ipaddress_enp0s8
    $consul_bootstrap_expect = '1'
}
```

Building a Single Container Application

Now, let's look at the parameters that we have set:

- `$docker_version`: This is the version of Docker that we will install.
- `$docker_tcp_bind`: This is the IP address and port that the Docker API will bind to.
- `$docker_image`: This is the Docker image we will be using from Docker Hub. We will be using my Consul image. To read more about the image or get the Dockerfile, go to https://hub.docker.com/r/scottyc/consul/.
- `$container_hostname`: This is going to set the hostname inside the container.
- `$consul_advertise`: This is the IP address that Consul is going to advertise. We are going to use a built-in Puppet fact, `$::ipaddress_enp0s8`.
- `$consul_bootstrap_expect`: This sets the number of nodes in the Consul cluster. We are using just the one. If it was a production cluster, you would use at least three.

Now that we have set up our parameters, let's get started on `install.pp`. As the name implies, this class will contain the logic that installs Docker, pulls the image, and runs the container. So, let's take a look at the code shown in the following screenshot:

```
class consul::install {

  package { 'device-mapper-libs':
    ensure => installed,
  }

  class { 'docker':
    version     => $consul::docker_version,
    tcp_bind    => $consul::docker_tcp_bind,
    socket_bind => 'unix:///var/run/docker.sock',
    require     => Package['device-mapper-libs']
  } ->

  docker::image { $consul::docker_image : } ->

  docker::run { $consul::container_hostname:
    image    => $consul::docker_image,
    hostname => $consul::container_hostname,
    command  => "-server --advertise ${consul::consul_advertise} -bootstrap-expect ${consul::consul_bootstrap_expect}",
    ports    => ['8301:8301', '8301:8301/udp', '8302:8302', '8302:8302/udp', '8400:8400', '8500:8500', '53:53/udp']
  }
}
```

To look at the code in more depth, we will break the class into two, the Docker installation and the container configuration. In the Docker installation, the first piece of code is a simple package type for `device-mapper-libs`. The reason we make sure that this package and its dependencies are installed is that it will be the storage drive that Docker will use to mount the container's filesystem.

Now, we move on to the Docker install. We start by declaring the `docker` class. For this class, we will set the Docker version, calling the parameters we set in `params.pp`, and the version of Docker that we are using is 1.9.1 (which is the latest at the time of writing this book). The next piece of configurations we will declare are the Docker API's TCP bind. Again, we will call our `params.pp` class and set the value to `tcp://127.0.0.1:4242`. This binds the API to listen to the localhost address on the TCP port `4242`.

The last value we will set to our Docker install is the Unix socket, which Docker will use. We will declare this without calling a parameter. The last piece of code makes sure that `device-mapper-libs` is installed before Docker, as it is a prerequisite to the Docker install:

```
package { 'device-mapper-libs':
    ensure => installed,
}

class { 'docker':
    version     => $consul::docker_version,
    tcp_bind    => $consul::docker_tcp_bind,
    socket_bind => 'unix:///var/run/docker.sock',
    require     => Package['device-mapper-libs']
} ->
```

Now that we have Docker installed, let's look at the code to build our Consul container. The first class that we call is `docker::image`. This will pull the image from Docker Hub before we call the `docker::run` class. In the `docker::run` class, we set navmar as the same value as the container's hostname. We will get that value from `params.pp` and it will be set to `consul`.

The next configuration we will set is the image. Now, this is different from calling `docker::image`. When you call `docker::image`, it pulls the image from Docker Hub to the local filesystem. When we set the image value in the `docker::run` class, it sets the value for the base image where the container will be deployed from. The value is set to `scottyc/consul`, and again we will get that value from `params.pp`. The `hostname` parameter is going to set the hostname inside the container.

Building a Single Container Application

Now we get to the the `resource` attribute that passes running configurations' parameters to the container. The `command` attribute is an arbitrary attribute that allows you to pass configurations to the container at boot. In this case, we are going to pass the boot configuration for the Consul setting as the server role, the IP address that the Consul application will bind to, and the number of servers that are there in the Consul's cluster. In the first case, all the values in the arguments that we are passing to the `command` attribute come from `params.pp`:

```
docker::image { $consul::docker_image : } ->

docker::run { $consul::container_hostname:
    image    => $consul::docker_image,
    hostname => $consul::container_hostname,
    command  => "-server --advertise ${consul::consul_advertise} -bootstrap-expect ${consul::consul_bootstrap_expect}",
    ports    => ['8301:8301', '8301:8301/udp', '8302:8302', '8302:8302/udp', '8400:8400', '8500:8500', '53:53/udp']
    }
}
```

Now, last but definitely not least, let's look at what our `init.pp` file contains. The first thing that you will note at the top after the main class declaration is the mapping of all our parameters to `params.pp`. The reason we do this is to set any sensible configurations or defaults in `params.pp` and any sensitive data we can overwrite the defaults with Hiera lookups. We will look at Hiera lookups in the next chapter. The last line of code includes our `consul::install` class, which we covered in the preceding section:

```
class consul (

    $docker_version          = $consul::params::docker_version,
    $docker_tcp_bind         = $consul::params::docker_tcp_bind,
    $docker_image            = $consul::params::docker_image,
    $container_hostname      = $consul::params::container_hostname,
    $consul_advertise        = $consul::params::consul_advertise,
    $consul_bootstrap_expect = $consul::params::consul_bootstrap_expect,

) inherits consul::params {

    include consul::install

}
```

Now, let's run our module.

Running our module

Now that we have written our module, I am sure we are all keen to run it; however, before we can do that, there is one more piece of configuration that we need to add to `servers.yml` and `default.pp`. First, we need to make sure that our module `consul` is located in `modules/consul`. The next step is to open our `servers.yml` file and add the following line at the bottom of the shell commands:

```
- { shell: cp /home/vagrant/node-01/modules/* -R /tmp/modules }
```

This will copy our module into the correct module path in the vagrant box. We also need to forward the Consul port so that we can hit the GUI. This is done by adding `- { guest: 8500, host: 8500 }` to the forwarded port's attribute. It should look as shown in the following screenshot:

```
box: puppetlabs/centos-7.0-64-puppet-enterprise
cpu: 1
ip: "172.17.8.101"
name: node-01
forward_ports:
    - { guest: 8500, host: 8500 }
ram: 2048
shell_commands:
    - { shell: 'yum install -y wget git lvm2 device-mapper-libs' }
    - { shell: '/opt/puppet/bin/gem install r10k && ln -s /opt/puppet/bin/r10k /usr/bin/r10k || true'}
    - { shell: 'cp /home/vagrant/node-01/Puppetfile /tmp && cd /tmp && r10k puppetfile install --verbose' }
    - { shell: cp /home/vagrant/node-01/modules/* -R /tmp/modules }
```

Now, let's open our `manifests` directory and edit `default.pp`. We just need to add our module to the node definition. You can do this by adding the `include consul` configuration, as shown in the following screenshot, and saving both files:

```
node 'node-01' {
    include consul
}
```

Building a Single Container Application

Let's head to our terminal, change the directory to the root of our Vagrant repo, and type the `vagrant up` command. Now, if this box is already running, issue a `vagrant destroy -f && vagrant up` command. The output should look as shown in this screenshot:

Even though we a have successful run Puppet, it can take a couple of minutes for the container to come up the first time, as it downloads the image from Docker Hub, so just be patient. You can easily check when the container is up by going to your browser and navigating to `127.0.0.1:8500`. You should get the consul GUI, as shown in the following screenshot:

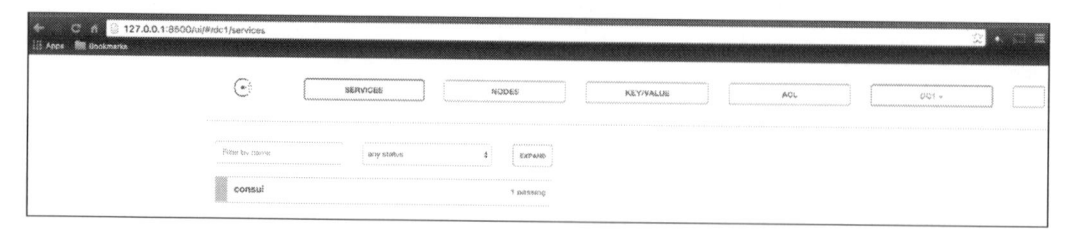

As you can see, we have one node that is running, named `consul`, which is the hostname that we gave to our container.

Coding using .erb files

In this topic, we are going to deploy the same container using `docker-compose` with a twist. The twist is with Puppet, where we can turn the `docker-compose.yml` file into `docker-compose.yml.erb`. This allows us to take advantage of all the tools Puppet gives us to manipulate template files. This is definitely the way I like to deploy containers with Puppet; however, I will let you decide for yourself the method you like best as we continue through the book.

Writing our module with Docker Compose

In the chapter, we are going to look at how to use `docker-compose` as an `.erb` template file. In this example, we are only deploying a single container, but when an application contains five or six containers with links, this way is much more efficient than using the standard manifest declarations.

So, we will take our `consul` module from the last topic and modify it now to use `docker-compose`. If you want to keep that module, just make a copy. First, we are not going to touch `init.pp` and `params.pp`—they will remain the same. Now, let's look at `install.pp`:

```puppet
class consul::install {

  package { 'device-mapper-libs':
    ensure => installed,
  }

  class { 'docker':
    version     => $consul::docker_version,
    tcp_bind    => $consul::docker_tcp_bind,
    socket_bind => 'unix:///var/run/docker.sock',
    require     => Package['device-mapper-libs']
  } ->

  file { '/root/docker-compose.yml':
    ensure  => file,
    content => template('consul/docker-compose.yml.erb'),
  } ->

  docker_compose { $consul::container_hostname :
    ensure => present,
    source => '/root',
    scale  => ['1']
  }
}
```

Building a Single Container Application

As you can see in the preceding screenshot, the top half of the class is exactly the same. However, we still install `device-mapper-libs` and declare the `docker` class in exactly the same way. The next attribute is different though; here, we call the file resource type. The reason is that this is the configuration that is used to place our `docker-compose` file on the local filesystem. You can see that we are declaring the contents with a template file located in the `templates` directory of the module. We will come back to this in a minute.

Now, let's look at the last resource type in `install.pp`. We are calling the `docker_compose` type because this is the resource type that will run the `docker-compose` command to bring up our container. Let's look at the attributes that we have configured. The first is `navmar`; this will set the name tag in Docker for the container. We are calling this value from `params.pp`, and it will be set to `consul`. `ensure` is a puppet meta parameter container which ensures that the container is always there.

If we want to delete the container, we would have to set this value to `absent`. The next attribute is `source`; this sets the folder where the `docker-compose` command can find the `docker-compose` file. We have set this to `root`. You could change the value to any folder on your system. The last attribute is `scale`. This tells `docker-compose` how many containers we want. The value is set to `1` in this instance. If we were deploying an nginx web farm, we might set the value to a figure such as 5. Now, let's get back to that template file. The first thing we need to do is create a folder called `templates` in the root of our `consul` module:

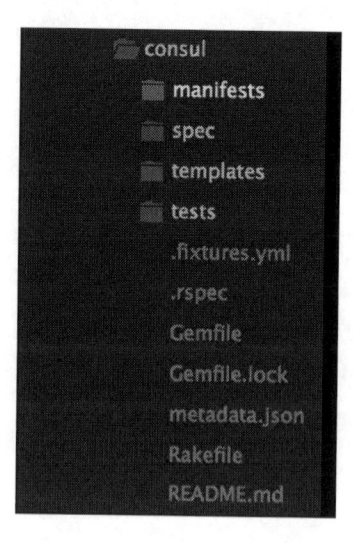

```
consul
    manifests
    spec
    templates
    tests
    .fixtures.yml
    .rspec
    Gemfile
    Gemfile.lock
    metadata.json
    Rakefile
    README.md
```

The next step after that is to create our `.erb` template file. In `install.pp`, we declared the filename as `docker-compose.yml.erb`, so in our `templates` directory, let's create a file with that name. The contents of the file should look as shown in the following screenshot:

```
<%= @container_hostname %>:
    image: <%= @docker_image %>
    hostname: <%= @container_hostname %>
    ports:
      - "8300:8300"
      - "8301:8301"
      - "8301:8301/udp"
      - "8302:8302"
      - "8302:8302/udp"
      - "8400:8400"
      - "8500:8500"
      - "53:53/udp"
    command: -server --advertise <%= @consul_advertise %> -bootstrap-expect <%= @consul_bootstrap_expect %>
```

So, the first thing that you should note in the preceding screenshot are the variables that are set, such as `<%= @container_hostname %>`. This maps back to `init.pp` as `$container_hostname`. As you can see, attributes such as `image`, `hostname`, `ports`, and `command` look very familiar. This is because they are the same attributes that we declared in the preceding section. In this example, we only configured our `docker-compose` file for a single container; in the next topic, we will look at a much more complex configuration. Before we get to that, let's make sure that this module runs.

Docker Compose up with Puppet

To run our module, let's make sure that our module is located in the `modules/consul` directory in the root of your Vagrant repo. We have the configuration to forward port 8500 (`forwarded_ports: - { guest: 8500, host: 8500 }`) and copy our module to our module's path directory using `- { shell: cp /home/vagrant/node-01/modules/* -R /tmp/modules }`.

Building a Single Container Application

Once this is in place, let's run `vagrant up` in our terminal in the root of our Vagrant repo. Again, if you have a box that is running, issue the `vagrant destroy -f && vagrant up` command. The terminal should give you the following output:

```
==> node-01: Running provisioner: puppet...
==> node-01: Running Puppet with default.pp...
==> node-01: Info: Loading facts
==> node-01: Info: Loading facts
==> node-01: Info: Loading facts
==> node-01: Notice: Compiled catalog for localhost in environment production in 1.29 seconds
==> node-01: Info: Applying configuration version '1453196319'
==> node-01: Notice: /Stage[main]/Docker::Repos/Yumrepo[docker]/ensure: created
==> node-01: Info: changing mode of /etc/yum.repos.d/docker.repo from 600 to 644
==> node-01: Notice: /Stage[main]/Docker::Install/Package[docker]/ensure: created
==> node-01: Notice: /Stage[main]/Docker::Service/File[/etc/sysconfig/docker-storage-setup]/ensure: created
==> node-01: Info: /Stage[main]/Docker::Service/File[/etc/sysconfig/docker-storage-setup]: Scheduling refresh of Service[docker]
==> node-01: Notice: /Stage[main]/Docker::Service/File[/etc/systemd/system/docker.service.d]/ensure: created
==> node-01: Notice: /Stage[main]/Docker::Service/File[/etc/systemd/system/docker.service.d/service-overrides.conf]/ensure: created
==> node-01: Info: /Stage[main]/Docker::Service/File[/etc/systemd/system/docker.service.d/service-overrides.conf]: Scheduling refresh of Exec[docker-systemd-reload]
==> node-01: Notice: /Stage[main]/Docker::Systemd_reload/Exec[docker-systemd-reload]: Triggered 'refresh' from 1 events
==> node-01: Notice: /Stage[main]/Docker::Service/File[/etc/sysconfig/docker]/ensure: created
==> node-01: Info: /Stage[main]/Docker::Service/File[/etc/sysconfig/docker]: Scheduling refresh of Service[docker]
==> node-01: Notice: /Stage[main]/Docker::Service/File[/etc/sysconfig/docker-storage]/ensure: created
==> node-01: Info: /Stage[main]/Docker::Service/File[/etc/sysconfig/docker-storage]: Scheduling refresh of Service[docker]
==> node-01: Notice: /Stage[main]/Docker::Service/Service[docker]/ensure: ensure changed 'stopped' to 'running'
==> node-01: Info: /Stage[main]/Docker::Service/Service[docker]: Unscheduling refresh on Service[docker]
==> node-01: Notice: /Stage[main]/Python/Package[python-dev]/ensure: created
==> node-01: Notice: /Stage[main]/Epel/File[/etc/pki/rpm-gpg/RPM-GPG-KEY-EPEL-7]/ensure: defined content as '{md5}58fa8ae27c89f37b08429f04fd4a88cc'
==> node-01: Notice: /Stage[main]/Consul::Install/File[/root/docker-compose.yml]/ensure: defined content as '{md5}8bdd827d303c79ce7f65826381517d13'
==> node-01: Notice: /Stage[main]/Epel/Epel::Rpm_gpg_key[EPEL-7]/Exec[import-EPEL-7]/returns: executed successfully
==> node-01: Notice: /Stage[main]/Epel/Yumrepo[epel-testing]/ensure: created
==> node-01: Info: changing mode of /etc/yum.repos.d/epel-testing.repo from 600 to 644
==> node-01: Notice: /Stage[main]/Epel/Yumrepo[epel-testing-debuginfo]/ensure: created
==> node-01: Info: changing mode of /etc/yum.repos.d/epel-testing-debuginfo.repo from 600 to 644
==> node-01: Notice: /Stage[main]/Epel/Yumrepo[epel-testing-source]/ensure: created
==> node-01: Info: changing mode of /etc/yum.repos.d/epel-testing-source.repo from 600 to 644
==> node-01: Notice: /Stage[main]/Epel/Yumrepo[epel]/ensure: created
==> node-01: Info: changing mode of /etc/yum.repos.d/epel.repo from 600 to 644
==> node-01: Notice: /Stage[main]/Epel/Yumrepo[epel-debuginfo]/ensure: created
==> node-01: Info: changing mode of /etc/yum.repos.d/epel-debuginfo.repo from 600 to 644
==> node-01: Notice: /Stage[main]/Epel/Yumrepo[epel-source]/ensure: created
==> node-01: Info: changing mode of /etc/yum.repos.d/epel-source.repo from 600 to 644
==> node-01: Notice: /Stage[main]/Python::Install/Package[pip]/ensure: created
==> node-01: Notice: /Stage[main]/Docker::Docker_compose/Python::Pip[docker-compose]/Exec[pip_install_docker-compose]/returns: executed successfully
==> node-01: Info: Checking if docker-compose.yml exists
==> node-01: Info: bring up containers
==> node-01: Notice: /Stage[main]/Consul::Install/Docker_compose[consul]/ensure: created
==> node-01: Notice: Finished catalog run in 75.34 seconds
```

Again, we can go to `127.0.0.1:8500` and get the Consul GUI:

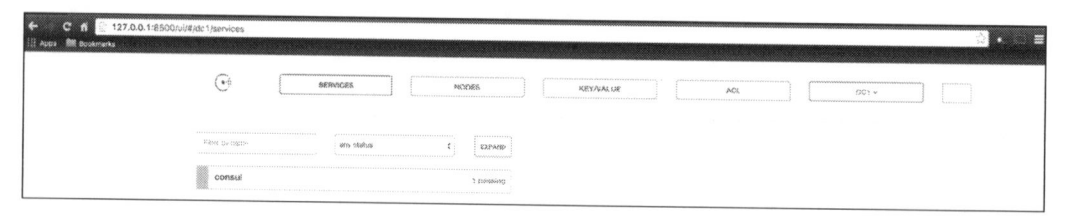

Now, let's log in to our vagrant box; we will do that by issuing the `vagrant ssh` command in our terminal from the root of our Vagrant repo. Once we have logged in, we can `su` to root (`sudo -i`). Then, we can issue the `docker ps` command to look at all the running containers. The terminal should give you the following output:

As you can see, the container is up and running.

Summary

In this chapter, we deployed our first container with Puppet. In doing so, we actually covered a lot of ground. We now have our script to create Puppet modules, and we know how to map our Puppet module's dependencies with `metadata.json` and `.fixtures.yml`.

In our tool belt, we now have two ways to deploy a container with Puppet, which will come in very handy in the chapters to come.

4
Building Multicontainer Applications

In the last few chapters, we built some cool stuff with Puppet and Docker. It has all been straight forward. Now, we are getting into the more complex topics such as how to keep a state and why to keep a state in a container. Docker is known to couple hand in hand with the micro service architecture, where the container would for the most part just be receiving data and transforming it into an output for a downstream application or data source to ingest. Thus, the container never keeps any state. Docker does not limit itself to such applications. In this chapter, we are going to prove this by building a fully functional Bitbucket server with a separate Postgres backend. We will mount the data that is stateful to the underlying host and decouple the applications from that state.

The other good feature of building Bitbucket is that you can use it for your Git server at home to make sure that all your future modules are in source control. Atlasssian offers a $10 dollar license for start-ups or developers that allow up to 10 users, which is a bargain.

In this chapter, we will build applications using standard manifest resources and then Docker Compose. Before we jump into coding, let's talk a little more about decoupling the state of an application from the actual application itself. The following are the topics that we will cover in this chapter:

- Decoupling a state
- Docker_bitbucket (manifest resources)
- Docker_bitbucket (Docker Compose)

Decoupling a state

In this topic, we are going to talk about state in containers. This topic will be all theory, but don't worry, we are going to put that theory to test when we write our module. We need to understand the theory behind state versus stateless so that in future, when writing your own Puppet modules to ship Docker, you can make the right design choice about states.

State versus stateless

So, we briefly touched on the topic of state versus stateless in the preceding section, but now, let's get into the nuts and bolts of it. So, let's look at two example applications: one with state and one without. First, we will start with a redis container. Its job is to be the backend store for activemq. The application that uses the queue has logic to check whether the message has been received and expects a response. If that fails, it will retry to send the message. Thus, the queue itself is stateless. We are using redis as a cache. There is no need to give the redis container state, as the information it holds is ephemeral. If the container fails, we just respawn a new one and it will cache the queue again.

Now, let's look at a container that needs state and the options we have to keep state. In the application, we are going to build two containers. One is Bitbucket itself, and then Postgres will be used as the backend database. So, both of these need state. Let's look at Bitbucket first. If we don't give states to the Bitbucket server, every time we restart the container, we would lose all the projects that we checked in via Git. That doesn't sound like it would be a viable solution. Now, let's look at the options that we have to give to the container's state. First, we could add volume to our Dockerfile:

```
EXPOSE 53/udp 8300 8301 8301/udp 8302
VOLUME ["/data"]
```

This will give the container state; we can reboot the application and all the data will be there, which is good. That is what we want. There is a downside to this method. Volume resides inside the container and is not decoupled from the application. So when we run into issues with this method, the main operational issue is when we need to update the version of the application in the container. As all our data lives in the container, we can't just pull the latest. So we are stuck with the available options. Now, let's look at how to map a folder from localhost to the container. We do this using `/data:/var/atlassian/application-data/bitbucket`. The left side of the colon is the localhost running the Docker daemon and the right side is the container.

Docker uses fully qualified paths, so it will create the `/data` directory. Now, we have decoupled our application from our data. If we want to update the version of Bitbucket, all we would need to do is change the `image` tag in our Puppet code to the new version. Then, we would run Puppet. Once that is complete, the new version of Bitbucket will boot with our existing data. Now, there is a cost to using this method as well. We have now tied this container to the one host. If you are using schedulers such as Kubernetes or Docker Swarm, this probably isn't the best idea. This problem has been solved by the new volume driver added to the Docker engine 1.8 and above. This allows us to create storage objects that are external of the host on which the engine is running.

> This is out of the scope of this book, but if you would like to do some more reading about the technology, I would recommend that you visit `https://clusterhq.com/flocker/introduction/`. Now, we have a good understanding of state versus stateless containers. Let's start the fun bit and get coding!

Docker_bitbucket (manifest resources)

In this topic, we are going to write our module that will install Atlassian's Bitbucket. The application will be comprised of two containers, among which one will be Postgres as we mentioned earlier. We are actually going to tweak the container to secure the backend and make it only accessible to Bitbucket. Then, we will run the Bitbucket container, configure the user it will run as, and then map the filesystem from the host machine to the container.

Creating our module skeleton

This is just going to be a quick refresher, as we covered this in the last chapter. If you are still not feeling comfortable with this step, I would recommend you to go over the last chapter again until you have a good grasp of the process.

So, we are going to open up our terminal and change directory or cd into the root of our Vagrant repo. Then, we will type `vagrant up`, and once the box is up, we will use SSH into it with `vagrant ssh`. The next step is to change to root (`sudo -i`). Now that we are root, let's change the directory to `/vagrant`, which maps back to our local machine. We are then going to issue the `puppet module generate <AUTHOR>-docker_bitbucket` command. Again, there are a few more tweaks that we need, but they are in the last chapter, so let's not repeat ourselves here. Once you have completed the outstanding task, you can move on to the next chapter.

Let's code

Now, we have our module skeleton and we have moved it into the `modules` directory in the root of our Vagrant environment. We need to add two new files: `install.pp` and `params.pp`. Our module should look like as shown in the following screenshot:

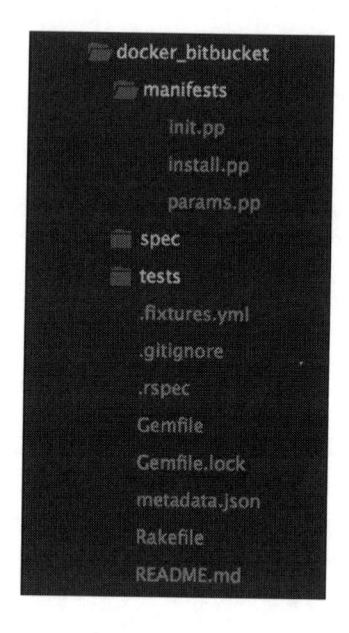

```
docker_bitbucket
    manifests
        init.pp
        install.pp
        params.pp
    spec
    tests
    .fixtures.yml
    .gitignore
    .rspec
    Gemfile
    Gemfile.lock
    metadata.json
    Rakefile
    README.md
```

In this example, we have a few new things going on, so I have not used `params.pp` in this example. This gives you a perfect opportunity to use the knowledge that you gained in the last chapter and apply it. So, for now, we will leave `params.pp` empty. Seeing as we are not putting parameters in `init.pp`, let's look at that first:

```
class docker_bitbucket {
  include docker_bitbucket::install
}
```

Chapter 4

As you can see in the preceding screenshot, we are only calling the `docker_bitbucket::install` class. We can now move on to the chunky `install.pp` class. Again, we are going to brake this down into three parts. This will make it easier to explain the logic of the class. Let's look at part one, which is as follows:

```puppet
package { 'device-mapper-libs':
    ensure => installed,
}

class { 'docker':
    version     => '1.9.1-1.el7.centos',
    tcp_bind    => 'tcp://127.0.0.1:4243',
    socket_bind => 'unix:///var/run/docker.sock',
    require     => Package['device-mapper-libs']
} ->
```

In this top section of the class, we are installing the `device-mapper-libs` package. This is a prerequisite for the RHEL family and Docker. The next thing that we are declaring is the Docker class. In this resource, we are defining the version of Docker that we want to be installed, the TCP bind that Docker will use, and lastly, the Unix socket that Docker will bind to. This is the same configuration that we used to define our Docker daemon in the last chapter. This will stay fairly static until we move into Docker schedulers. Let's now move to Postgres:

```puppet
docker::image { 'postgres:9.2': } ->

docker::run { 'postgres':
    image    => 'postgres:9.2',
    hostname => 'bitbucket-db',
    env      => ['POSTGRES_USER=postgresql', 'POSTGRES_PASSWORD=Gr33nTe@', 'POSTGRES_DB=bitbucket', 'PGDATA=/var/lib/postgresql/data/pgdata'],
    volumes  => ['/root/db:/var/lib/postgresql/data/pgdata']
}
```

First, we will define the image of Postgres that we would like to use. For this example, we are using Postgres 9.2. So, the correct tag from Docker Hub is `postgres:9.2`. Now, let's look at the `docker::run` class; this is where all the configurations for Postgres will be defined. So, you can see that we are calling the image that we set in the preceding resource `postgres:9.2`. We will then set the hostname as `bitbucket-db`. This setting is important, so let's store it into our memory for later use.

Let's look at the `env` resource declaration as we have a bit going on there. In this one line, we are declaring the Postgres user, the database password, the name of the database that we will connect with Bitbucket, and lastly the path where Postgres with store the database. Lastly, we are declaring our volumes as `/root/db:/var/lib/postgresql/data/pgdata`.

Building Multicontainer Applications

As mentioned earlier, the left-hand side of the colon is mapping the the local machine and the right is mapping the container. There are two major call outs with our configuration. First, the `/root/db` folder is arbitrary and not what you would use in production. The second is that you will note that the left side of the colon, `/var/lib/postgresql/data/pgdata`, and the value in env, `PGDATA`, are the same. This is no coincidence; that folder holds the only state that we care about: the actual database. That is the only state that we will keep. Nothing more and nothing less. You will note that we have not exposed any ports from this container. This is by design. We are going to link our Bitbucket container to Postgres. What does link mean? It means that by default, the Postgres image exposes port 5432. This is the port that we will use to connect to our database. By linking the containers, only the Bitbucket container has access to 5432; if we exposed the port (5432:5432), any other application that has access to the host instance could hit the port. Thus, linking is much more secure. So, we need to remember a few things from this section of code for later use: the hostname and the entire env line. For now, let's move on to the Bitbucket container:

```
docker::image { 'atlassian/bitbucket-server': } ->

docker::run { 'bitbucket':
    image    => 'atlassian/bitbucket-server',
    ports    => ['7990:7990', '7999:7999'],
    username => 'root',
    volumes  => ['/data:/var/atlassian/application-data/bitbucket'],
    links    => ['postgres']
}
```

As you can see in the preceding screenshot, the image resources are the same, but instead of calling Postgres, we are going to call `atlassian/bitbucket-server`. The next resource we will declare is the ports resource. You will note that we are declaring two ports `7990:7990`, which will be the port that we hit the web UI on, and `7999:7999`, which is the port that Bitbucket uses for SSH. We will set the username to `root`. This is recommended in Atlassian's documentation (https://hub.docker.com/r/atlassian/bitbucket-server/).

Next, we are going to map our volume drive. In this case, we are only going to map Bitbucket's data directory. This is where all our Git repo, user information, and so on is kept. Again, `/data` is an arbitrary location; you could use any location you like. The important location to note is on the left-hand side of the colon, `/var/atlassian/application-data/bitbucket`.

Lastly, we link our two containers. Another benefit of linking containers is that the Docker daemon will write to both the containers' `/etc/hosts` files with their hostname and IP address. So, the containers have no issue talking to each other. There is no need to worry about the IP address, as it is arbitrary and is looked after by the Docker daemon. Now that we have written our module, we can build our application.

Running our module

The first thing that we need to do is forward the correct ports on our `servers.yml` file that allow us to hit the ports we forwarded to Bitbucket. To do this, we need to modify the file so that it looks as shown in the following screenshot:

```
---
-
  box: puppetlabs/centos-7.0-64-puppet-enterprise
  cpus: 1
  ip: "172.17.8.101"
  name: node-01
  forward_ports:
    - { guest: 7990, host: 7990 }
    - { guest: 7999, host: 7999 }
  ram: 2048
  shell_commands:
    - { shell: 'yum install -y wget git lvm2 device-mapper-libs' }
    - { shell: '/opt/puppet/bin/gem install r10k && ln -s /opt/puppet/bin/r10k /usr/bin/r10k || true' }
    - { shell: 'cp /home/vagrant/node-01/Puppetfile /tmp && cd /tmp && r10k puppetfile install --verbose' }
    - { shell: cp /home/vagrant/node-01/modules/* -R /tmp/modules }
```

So, let's open our terminal and change the directory to the root of our Vagrant repo and run `vagrant up`. You should get the following output after this:

Building Multicontainer Applications

Now that our application is built, we can go to `http://127.0.0.1:7990`. We should get the following page:

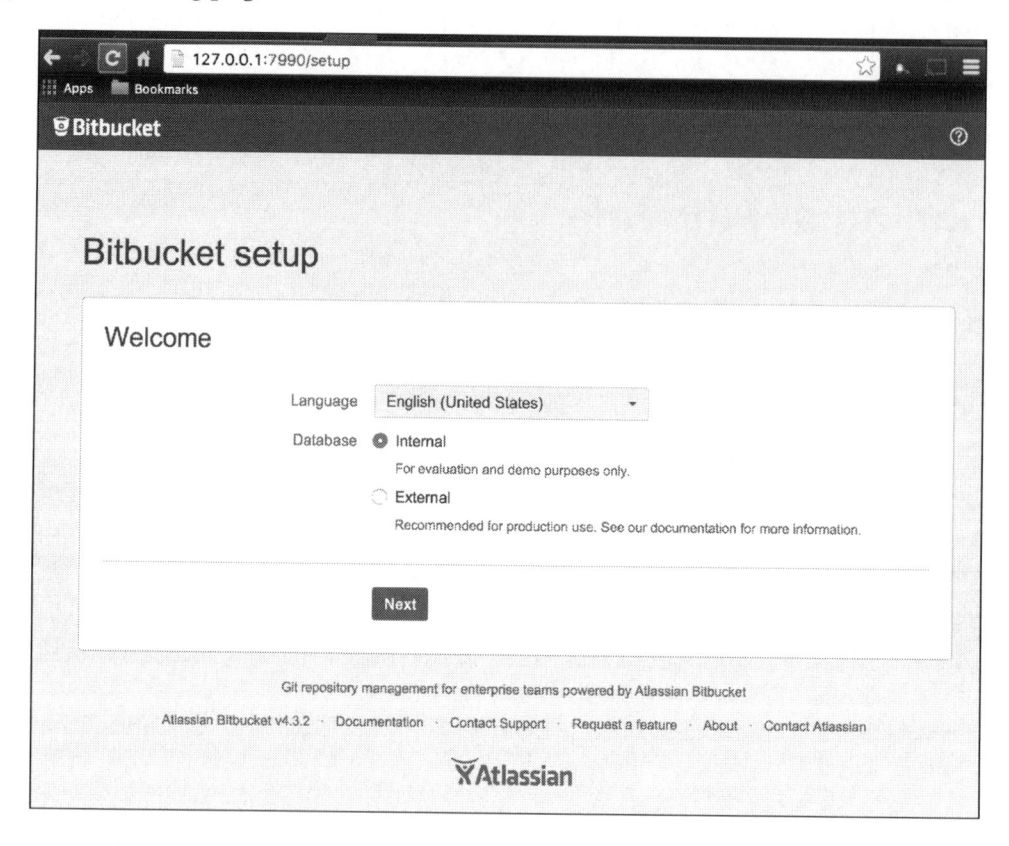

Earlier in the topic, we had remembered some details about our Postgres install. Now is the time to use them, so let's begin. The first thing that we need to do is use an external database. The next piece of configuration we need to choose is the database type. Of course, we are going to choose Postgres.

The hostname will be set to the hostname of the `bitbucket-db` container, the port is `5432`, and the database name is `bitbucket`, as we set in our code. We will use PostgreSQL as the username and the password will be `Gr33nTe@`. Refer to the following screenshot to know more:

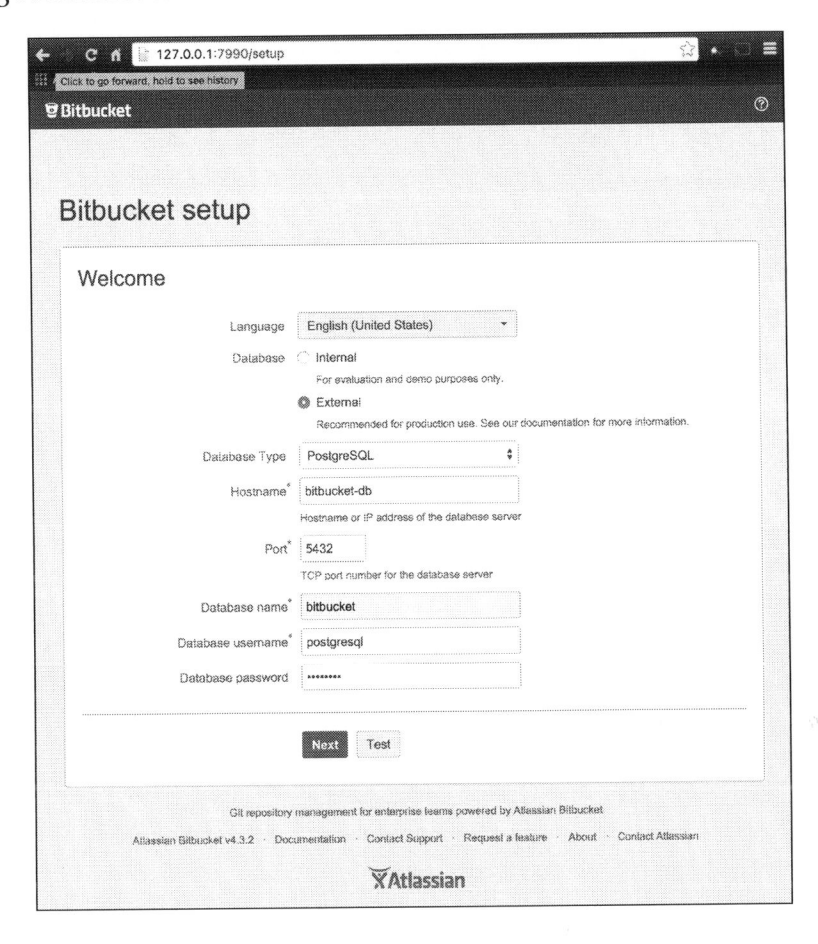

Building Multicontainer Applications

Next, click on the **Test** button, and we should get the **Successfully established database connection.** message, as shown in the following screenshot:

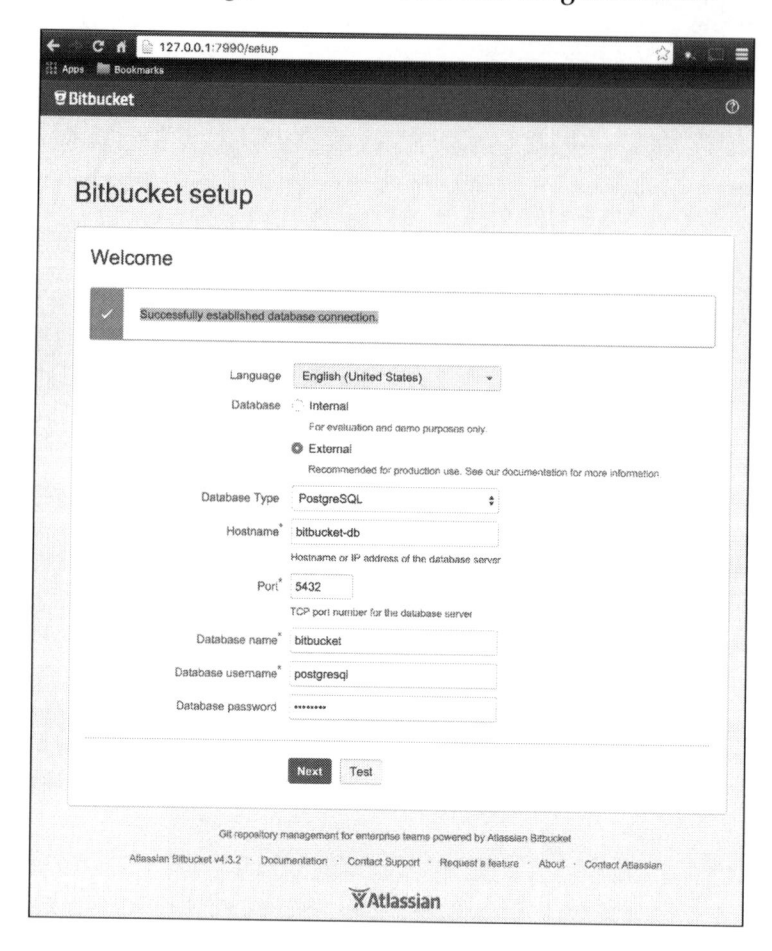

I will let you finish the rest of the setup. But what we just set up was not simple, and now, we have a very solid base to move on to more complex applications.

Docker_bitbucket (Docker Compose)

In this topic, we are going to build the same Bitbucket application. The difference this time is that we are going to use `docker-compose` as a `.erb` file instead of the resource declarations in a manifest.

Let's code – take 2

We covered a lot of what happens under the hood in the last topic. We will not be repeating ourselves, so this topic will be just about the code. We are going to keep both `init.pp` and `params.pp` the same as we did in the last topic. So, let's jump straight to `install.pp`. It will look very similar to `install.pp` from the last chapter:

```puppet
class docker_bitbucket::install {

  package { 'device-mapper-libs':
    ensure => installed,
  }

  class { 'docker':
    version     => '1.9.1-1.el7.centos',
    tcp_bind    => 'tcp://127.0.0.1:4243',
    socket_bind => 'unix:///var/run/docker.sock',
    require     => Package['device-mapper-libs']
  } ->

  file { '/root/docker-compose.yml':
    ensure  => file,
    content => template('docker_bitbucket/docker-compose.yml.erb'),
  } ->

  docker_compose { 'bitbucket' :
    ensure => present,
    source => '/root',
    scale  => ['1']
  }
}
```

All the magic happens in our template file. So, let's jump to our `.erb` file that lives in the `templates` folder in the root of our module:

```yaml
postgres:
    image: postgres:9.2
    environment:
        - POSTGRES_USER=postgresql
        - POSTGRES_PASSWORD=Gr33nTe@
        - POSTGRES_DB=bitbucket
        - PGDATA=/var/lib/postgresql/data/pgdata
    volumes:
        - ./db:/var/lib/postgresql/data/pgdata
bitbucket:
    image: atlassian/bitbucket-server
    user: root
    ports:
        - "7990:7990"
        - "7999:7999"
    volumes:
        - ./data:/var/atlassian/application-data/bitbucket atlassian/bitbucket-server chown -R daemon /var/atlassian/application-data/bitbucket
        - ./data:/var/atlassian/application-data/bitbucket
    links:
        - postgres
```

As you can see in our `.erb` file in the preceding screenshot, all the configurations are familiar. There are absolutely no changes to what we covered in our last topic.

Building Multicontainer Applications

Running our module – take 2

Let's open our terminal and change the directory to the root of our Vagrant repo and run `vagrant up`. You should get the following output:

Now, let's just go to `http://127.0.0.1:7990`, and we should get the following page:

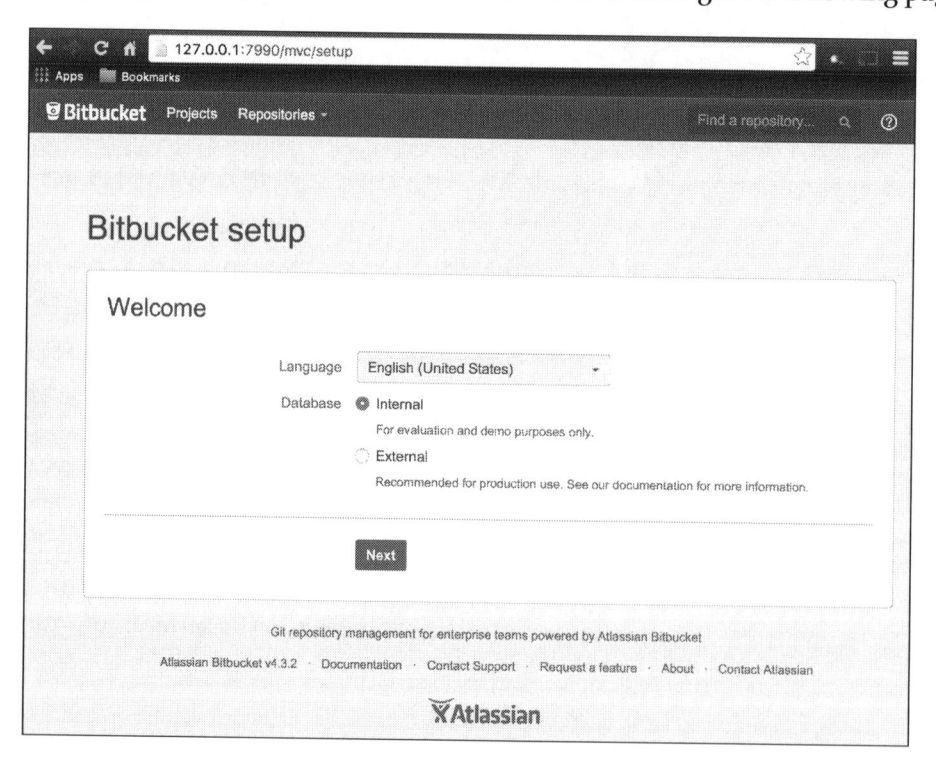

Just follow the same setup as in the preceding topic to configure Bitbucket. You can use a trail license to try the application, or as I mentioned earlier, there is a development/startup license at `https://bitbucket.org/product/pricing?tab=server-pricing` with the proceeds of the $10 license going to charity.

Summary

By building a multicontainer application, we learned and covered a lot. We first looked at state versus stateless containers, the pros and cons of having state, and what design choices we have to keep a state. We then looked at linked containers and how they communicate with each other through their hostfiles. All the topics in this chapter will set us up with the knowledge that we need to move forward with topics such as service discovery and container schedulers.

Configuring Service Discovery and Docker Networking

In this chapter, we will be looking at two very important topics when working with containers. First, we will be looking at what is service discovery, why do we need it, and the different types of service discovery. The second topic we will cover is Docker networking. There are many ways to run container networks. There are some great technologies out there such as the CoreOS project flannel (https://coreos.com/flannel/docs/latest/). There is also Weave from Weave Works (http://weave.works/), but we are going to use the native Docker networking stack released in engine version 1.9.1.

Service discovery

This is a fairly important topic in the world of containers, when we start to move into multinode applications and Docker schedulers. The question is what is service discovery? Is it limited to containers? What are the types of service discovery for us to make smart design choices in our Puppet modules.

The theory

Service discovery is essential when we start to work with multinode applications, as it allows our applications to talk to each other as they move from node to node. So, as you can see in the world of containers, this is fairly important. We have a few choices when we choose a service discovery backend. The two big names in this space are **etcd** (https://coreos.com/etcd/), which again is from CoreOS, and **Consul** from HashiCorp (https://www.consul.io/).

You might remember that we have already written a `consul` module. So for this chapter, we are going to choose the same, as we already have the written code. First, let's look at the architecture of Consul so we can understand how the backend works, how it handles failure, and what option do we get with our configuration of Consul.

So, let's talk about how Consul works. In Consul, we have two types of configuration that we can give to a server. First is a server role and the second is an agent role. Although the two interact, they serve different purposes. Lets dive into the server role first. The server's role is to participate in the RAFT quorum; this is to maintain the state of the cluster. In Consul, we have the idea of data centers. What is a data centre, you may ask? It is a group of logical servers and agents. For example, if you are an AWS, a data center could be an AZ or even a VPC. Consul allows connectivity between data centers; it is the role of the sever to look after the communications between data centers. Consul uses the gossip protocol to achieve this. The server also holds the key/value store and replicates it between the servers using the serf protocol. Let's look at a diagram of what we discussed:

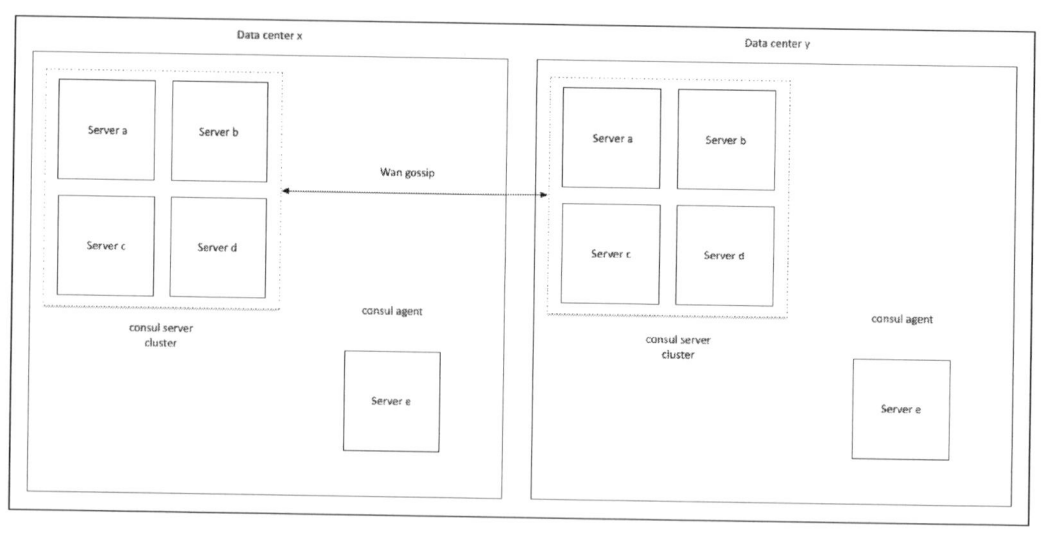

The agent's role is to report to the server about the state of the machine and any health checks that may be assigned to it. Again, Consul will use the serf protocol to pass the communication.

Now that we have an understanding of what Consul is doing behind the scenes, let's look at the features that it has that we can take advantage of in our Puppet modules. The first feature we will take advantage of is DNS service discovery. In the container world, this is pretty important. As our containers move from node to node, we need to know how to connect to them. DNS service discovery solves this very neatly. So, let's look at an example to understand this.

Chapter 5

In this example, we have a **mario** service and we have **Docker swarm cluster** of three nodes. When we hit the Docker API and swarm schedules the container, we don't know which of the three machines **mario** will end up on. But we have other services that will need to find **mario** as soon as he is up. If we tell the other services that **mario** is actually at **mario.service.consul**, no matter what node the container comes up on, it will resolve **mario.service.consul** to the right address. Refer to the following diagram to understand this in detail:

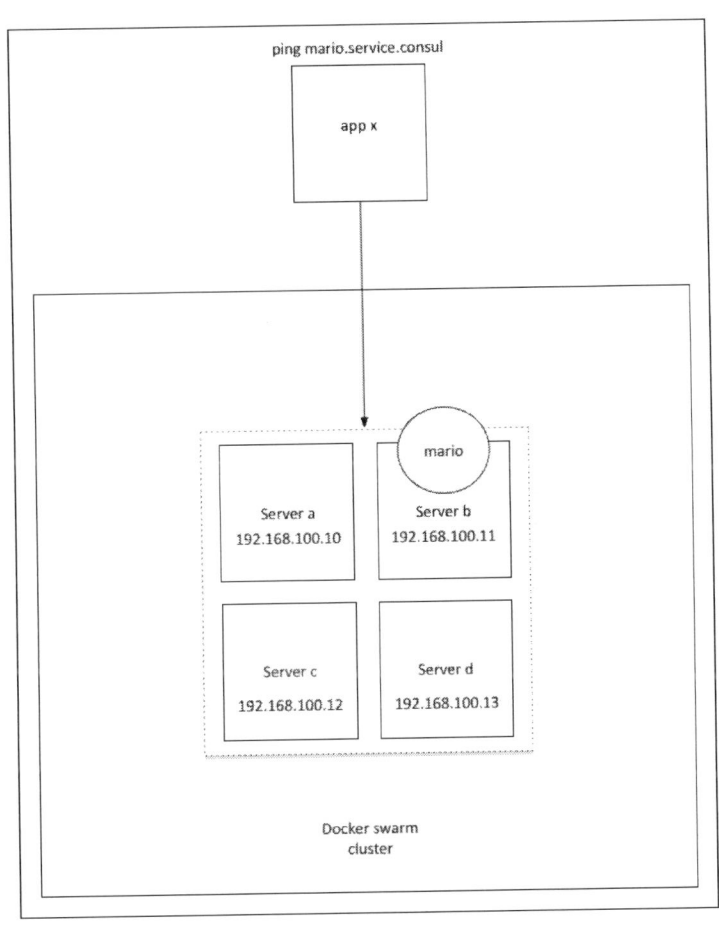

In this case, if we were to ping **mario.service.consul**, we would get **192.168.100.11**. Pending our scheduling configuration in swarm, if **Server b** fails, **mario.service.consul** could end up on **Server d**. So, the response to **mario.service.consul** would now come from **192.168.100.13**. This would take no human intervention and would be seamless to the applications. That is all the theory we will see for service discovery in this chapter; there is more that we will cover in the later chapters. Now, let's get to writing some code.

The service discovery module

In this module, we are going to write a module that uses consul as our DNS service discovery backend. As we already have a consul module, we won't start from scratch but add new features to the existing module. We will again write the module with manifests and Docker Compose. So, let's start with the manifests.

Our folder structure should look like this:

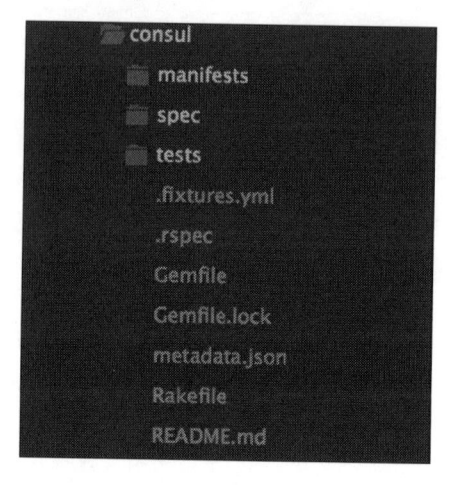

```
consul
    manifests
    spec
    tests
    .fixtures.yml
    .rspec
    Gemfile
    Gemfile.lock
    metadata.json
    Rakefile
    README.md
```

Let's jump straight to `install.pp`. Without making any changes, it should look as shown in the following screenshot:

```puppet
class consul::install {

  package { 'device-mapper-libs':
    ensure => installed,
  }

  class { 'docker':
    version     => $consul::docker_version,
    tcp_bind    => $consul::docker_tcp_bind,
    socket_bind => 'unix:///var/run/docker.sock',
    require     => Package['device-mapper-libs']
  } ->

  docker::image { $consul::docker_image : } ->

  docker::run { $consul::container_hostname:
    image    => $consul::docker_image,
    hostname => $consul::container_hostname,
    command  => "-server --advertise ${consul::consul_advertise} -bootstrap-expect ${consul::consul_bootstrap_expect}",
    ports    => ['8301:8301', '8301:8301/udp', '8302:8302', '8302:8302/udp', '8400:8400', '8500:8500', '53:53/udp']
  }
}
```

Now, we are going to add one extra container that is going to be part of the plumbing for our DNS service discovery solution. We will need something to register our containers with Consul as they spawn. For this, we will use a golang application called **registrator** (https://github.com/gliderlabs/registrator). This is a fantastic app. I have been using it for over a year, and it has been faultless. So, let's make changes to our `params.pp` file to allow the new container. At the moment, `params.pp` looks like the one shown in the following screenshot:

```
1    class consul::params {
2
3        $docker_version            = '1.9.1-1.el7.centos'
4        $docker_tcp_bind           = 'tcp://127.0.0.1:4243'
5        $docker_image              = 'scottyc/consul'
6        $container_hostname        = 'consul'
7        $consul_advertise          = $::ipaddress_enp0s8
8        $consul_bootstrap_expect   = '1'
9    }
10
```

The first thing that we will do is make changes to the `docker_image` and `container_hostname` parameters. As we already have the convention of `consul_xxx`, we can carry on with that:

```
1    class consul::params {
2
3        $docker_version             = '1.9.1-1.el7.centos'
4        $docker_tcp_bind            = 'tcp://127.0.0.1:4243'
5        $consul_docker_image        = 'scottyc/consul'
6        $consul_container_hostname  = 'consul'
7        $consul_advertise           = $::ipaddress_enp0s8
8        $consul_bootstrap_expect    = '1'
9    }
10
```

Now, let's add the parameters for registrator:

```
1    class consul::params {
2
3        $docker_version             = '1.9.1-1.el7.centos'
4        $docker_tcp_bind            = 'tcp://127.0.0.1:4243'
5        $consul_docker_image        = 'scottyc/consul'
6        $consul_container_hostname  = 'consul'
7        $consul_advertise           = $::ipaddress_enp0s8
8        $consul_bootstrap_expect    = '1'
9        $reg_docker_image           = 'gliderlabs/registrator'
10       $reg_container_hostname     = 'registrator'
11       $reg_net                    = 'host'
12       $reg_volume                 = ['/var/run/docker.sock:/tmp/docker.sock']
13       $reg_command                = "consul://$::ipaddress_enp0s8:8500"
14   }
15
```

As you can see, we have added the parameter for the image as `$reg_docker_image = 'gliderlabs/registrator'` and the parameter for the hostname as `$reg_container_hostname = 'registrator'`. We have told the container to listen to the host's `$reg_net = 'host'` network. The next parameter will need some explaining. The registrator maps the Unix socket that the Docker daemon is bound to into its Unix socket. It does this to listen to any new services that get spawned and need to be registered in consul for discovery. As you can see, we do this with `$reg_volume = ['/var/run/docker.sock:/tmp/docker.sock']`. The last parameter tells registrator where to find `consul`. We are going to set that with `$reg_command = "consul://$::ipaddress_enp0s8:8500"`. Now, let's move over to our `init.pp` file.

Our `init.pp` file should look as shown in the following screenshot:

```
$docker_version              = $consul::params::docker_version,
$docker_tcp_bind             = $consul::params::docker_tcp_bind,
$consul_docker_image         = $consul::params::consul_docker_image,
$consul_container_hostname   = $consul::params::consul_container_hostname,
$consul_advertise            = $consul::params::consul_advertise,
$consul_bootstrap_expect     = $consul::params::consul_bootstrap_expect,
```

Let's add our new parameters, as shown in the following screenshot:

```
37  #
38  class consul {
39
40      $docker_version              = $consul::params::docker_version,
41      $docker_tcp_bind             = $consul::params::docker_tcp_bind,
42      $consul_docker_image         = $consul::params::consul_docker_image,
43      $consul_container_hostname   = $consul::params::consul_container_hostname,
44      $consul_advertise            = $consul::params::consul_advertise,
45      $consul_bootstrap_expect     = $consul::params::consul_bootstrap_expect,
46      $reg_docker_image            = $consul::params::reg_docker_image,
47      $reg_container_hostname      = $consul::params::reg_container_hostname,
48      $reg_net                     = $consul::params::reg_net,
49      $reg_volume                  = $consul::params::reg_volume,
50      $reg_command                 = $consul::params::reg_command,
```

Now that we have all our parameters set up, we can go to our `install.pp` file to add our code in order to install registrator:

```puppet
    docker::image { $consul::reg_docker_image : } ->

    docker::run { $consul::reg_container_hostname:
        image   => $consul::reg_docker_image,
        net     => $consul::reg_net,
        volumes => $consul::reg_volume,
        command => $consul::reg_command,
    }
}
```

As you can see in the preceding screenshot, we have added a new block of code at the bottom of our file. It's similar to our code that configures Consul; however, there are a few different parameters. We covered those earlier, so let's not repeat ourselves. Now that we've made a fair chunk of changes to our module, we should run it in Vagrant to check whether we have any issues. Before we can run Vagrant, we need to change our `servers.yaml` file in the root of our Vagrant repo so that it allows us to hit the Consul URL on port `8500`. We do this with the following change to the code:

```yaml
---
-
    box: puppetlabs/centos-7.0-64-puppet-enterprise
    cpu: 1
    ip: "172.17.8.101"
    name: node-01
    forward_ports:
        - { guest: 8500, host: 8500 }
    ram: 2048
    shell_commands:
        - { shell: 'yum install -y wget git lvm2 device-mapper-libs' }
        - { shell: '/opt/puppet/bin/gem install r10k && ln -s /opt/puppet/bin/r10k /usr/bin/r10k || true'}
        - { shell: 'cp /home/vagrant/node-01/Puppetfile /tmp && cd /tmp && r10k puppetfile install --verbose' }
        - { shell: cp /home/vagrant/node-01/modules/* -R /tmp/modules }
```

Configuring Service Discovery and Docker Networking

Now, let's open our terminal and change the directory to the root of our Vagrant repo. From there, we will just issue the `vagrant up` command. The output from our terminal should look as shown in the following screenshot:

After this, let's open our browser and go to `127.0.0.1:8500`:

You will notice now that there are a lot more services listed in the Consul web UI than when we ran the module in the last chapter. This is because now, registrator is listening on the Unix socket, and any container with a port mapped to the host will be registered. So the good news is that our module is working. Now, let's add an application to the module.

The easiest way to do this is to add another container module to our node. So, let's add our `bitbucket` module. We do this by adding the class to our `default.pp` file that lives in our `manifests` directory:

```
1  node 'node-01' {
2
3      include consul
4      include docker_bitbucket
5
6  }
7
8
```

We will also need to make some quick modifications to the `bitbucket` module so that we don't get duplicate declaration errors. Note that this is not something you would do in production. But it is good enough for our test lab. We need to comment out the top block of code as shown in the following screenshot:

```
1   class docker_bitbucket::install {
2
3   # package { 'device-mapper-libs':
4   #   ensure => installed,
5   # }
6
7   # class { 'docker':
8   #   version  => '1.8.1-1.el7.centos',
9   #   tcp_bind => 'tcp://127.0.0.1:4243',
10  # }
11
12
13
14  docker::image { 'postgres:9.2': } ->
15
16  docker::run { 'postgres':
17      image    => 'postgres:9.2',
18      hostname => 'bitbucket-db',
19      env      => ['POSTGRES_USER=postgresql', 'POSTGRES_PASSWORD=Gr33nTe@', 'POSTGRES_DB=bitbucket', 'PGDATA=/var/lib/postgresql/data/pgdata'],
20      volumes  => ['/root/db:/var/lib/postgresql/data/pgdata']
21  }
22
23  docker::image { 'atlassian/bitbucket-server': } ->
24
25  docker::run { 'bitbucket':
26      image    => 'atlassian/bitbucket-server',
27      username => 'root',
28      ports    => ['7990:7990', '7999:7999'],
29      volumes  => ['/data:/var/atlassian/application-data/bitbucket'],
30      links    => ['postgres']
31  }
32  }
33
34
35
```

[83]

Configuring Service Discovery and Docker Networking

We can even comment out the code as shown in the following screenshot:

```
{  1    class docker_bitbucket::install {
   2
   3
   4
   5
   6
   7
   8
   9
  10
  11
  12
  13
  14    file { '/root/docker-compose.yml':
  15      ensure   => file,
  16      content  => template('docker_bitbucket/docker-compose.yml.erb'),
  17    } ->
  18
  19    docker_compose { 'bitbucket' :
  20      ensure => present,
  21      source => '/root',
  22      scale  => ['1']
  23    }
} 24  }
  25
  26
  27
```

This depends on whether you used the `manifest` module or the `compose` module. I used the `compose` module.

So, let's go back to our terminal in the root of our Vagrant repo and issue the `vagrant provision` command. The output of the terminal should look as shown in the following screenshot:

==> node-01: Running Puppet with default.pp...
==> node-01: Info: Loading facts
==> node-01: Info: Loading facts
==> node-01: Info: Loading facts
==> node-01: Notice: Compiled catalog for localhost in environment production in 1.18 seconds
==> node-01: Info: Applying configuration version '1456398403'
==> node-01: Notice: /Stage[main]/Docker::Repos/Yumrepo[docker]/ensure: created
==> node-01: Info: changing mode of /etc/yum.repos.d/docker.repo from 600 to 644
==> node-01: Notice: /Stage[main]/Docker::Install/Package[docker]/ensure: created
==> node-01: Notice: /Stage[main]/Docker::Service/File[/etc/sysconfig/docker-storage-setup]/ensure: created
==> node-01: Info: /Stage[main]/Docker::Service/File[/etc/sysconfig/docker-storage-setup]: Scheduling refresh of Service[docker]
==> node-01: Notice: /Stage[main]/Docker::Service/File[/etc/systemd/system/docker.service.d]/ensure: created
==> node-01: Notice: /Stage[main]/Docker::Service/File[/etc/systemd/system/docker.service.d/service-overrides.conf]/ensure: created
==> node-01: Info: /Stage[main]/Docker::Service/File[/etc/systemd/system/docker.service.d/service-overrides.conf]: Scheduling refresh of Exec[docker-systemd-reload-before-service]
==> node-01: Notice: /Stage[main]/Docker::Service/Exec[docker-systemd-reload-before-service]/returns: executed successfully
==> node-01: Notice: /Stage[main]/Docker::Service/Exec[docker-systemd-reload-before-service]: Triggered 'refresh' from 1 events
==> node-01: Notice: /Stage[main]/Docker::Service/File[/etc/sysconfig/docker-storage]/ensure: created
==> node-01: Info: /Stage[main]/Docker::Service/File[/etc/sysconfig/docker-storage]: Scheduling refresh of Service[docker]
==> node-01: Notice: /Stage[main]/Docker::Service/File[/etc/sysconfig/docker]/ensure: created
==> node-01: Info: /Stage[main]/Docker::Service/File[/etc/sysconfig/docker]: Scheduling refresh of Service[docker]
==> node-01: Notice: /Stage[main]/Docker::Service/Service[docker]/ensure: ensure changed 'stopped' to 'running'
==> node-01: Info: /Stage[main]/Docker::Service/Service[docker]: Unscheduling refresh on Service[docker]
==> node-01: Notice: /Stage[main]/Docker::Compose/Exec[Install Docker Compose 1.6.0]/returns: executed successfully
==> node-01: Notice: /Stage[main]/Docker::Compose/File[/usr/local/bin/docker-compose-1.6.0]/mode: mode changed '0644' to '0755'
==> node-01: Notice: /Stage[main]/Docker::Compose/File[/usr/local/bin/docker-compose]/ensure: created
==> node-01: Notice: /Stage[main]/Docker_bitbucket::Install/File[/root/docker-compose.yml]/ensure: defined content as '{md5}c6c19b0c5b81776d46481734d67c610f'
==> node-01: Info: Checking if docker-compose.yml exists
==> node-01: Info: bring up containers
==> node-01: Notice: /Stage[main]/Docker_bitbucket::Install/Docker_compose[bitbucket]/ensure: created
==> node-01: Notice: /Stage[main]/Consul::Install/Docker::Image[scottyc/consul]/File[/usr/local/bin/update_docker_image.sh]/ensure: created
==> node-01: Notice: /Stage[main]/Consul::Install/Docker::Image[scottyc/consul]/Exec[/usr/local/bin/update_docker_image.sh scottyc/consul]/returns: executed successfully
==> node-01: Notice: /Stage[main]/Consul::Install/Docker::Image[gliderlabs/registrator]/Exec[/usr/local/bin/update_docker_image.sh gliderlabs/registrator]/returns: executed successfully
==> node-01: Notice: /Stage[main]/Consul::Install/Docker::Run[consul]/File[/etc/systemd/system/docker-consul.service]/ensure: created
==> node-01: Info: /Stage[main]/Consul::Install/Docker::Run[consul]/File[/etc/systemd/system/docker-consul.service]: Scheduling refresh of Exec[docker-systemd-reload]
==> node-01: Info: /Stage[main]/Consul::Install/Docker::Run[consul]/File[/etc/systemd/system/docker-consul.service]: Scheduling refresh of Service[docker-consul]
==> node-01: Notice: /Stage[main]/Consul::Install/Docker::Run[registrator]/File[/etc/systemd/system/docker-registrator.service]/ensure: created
==> node-01: Info: /Stage[main]/Consul::Install/Docker::Run[registrator]/File[/etc/systemd/system/docker-registrator.service]: Scheduling refresh of Exec[docker-systemd-reload]
==> node-01: Info: /Stage[main]/Consul::Install/Docker::Run[registrator]/File[/etc/systemd/system/docker-registrator.service]: Scheduling refresh of Service[docker-registrator]
==> node-01: Notice: /Stage[main]/Docker::Systemd_reload/Exec[docker-systemd-reload]: Triggered 'refresh' from 2 events
==> node-01: Notice: /Stage[main]/Consul::Install/Docker::Run[consul]/Service[docker-consul]/ensure: ensure changed 'stopped' to 'running'
==> node-01: Info: /Stage[main]/Consul::Install/Docker::Run[consul]/Service[docker-consul]: Unscheduling refresh on Service[docker-consul]
==> node-01: Notice: /Stage[main]/Consul::Install/Docker::Run[registrator]/Service[docker-registrator]/ensure: ensure changed 'stopped' to 'running'
==> node-01: Info: /Stage[main]/Consul::Install/Docker::Run[registrator]/Service[docker-registrator]: Unscheduling refresh on Service[docker-registrator]
==> node-01: Notice: Finished catalog run in 245.82 seconds

Now, let's look at our browser again. We can see that our `bitbucket` services have been registered, as shown in this screenshot:

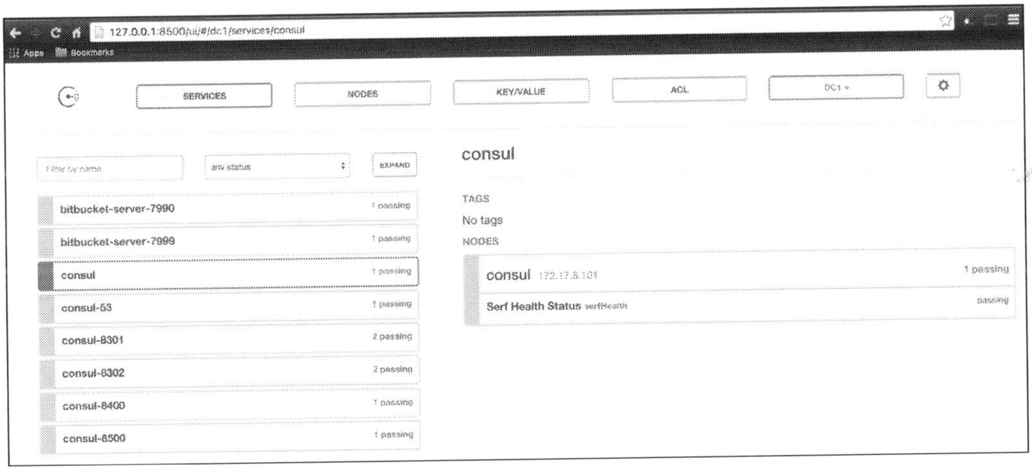
127.0.0.1:8500/ui/#/dc1/services/consul
SERVICES
NODES
KEY/VALUE
ACL
consul
TAGS
No tags
NODES
bitbucket-server-7990
bitbucket-server-7999
consul
consul-53
consul-8301
consul-8302
consul-8400
consul-8500
Serf Health Status

We have the service discovery working; however, we still need to add another class to our module for DNS service discovery. So, let's go back to our `consul` module. We will add a new file called `package.pp`. In this file, we will install the bind package and add two templates, one to configure `named.conf` and the other to configure `consul.conf` in the directory named `/etc/`. Let's start coding. The first thing we will need to do is create our `package.pp` file in the `manifests` directory of our module:

```
consul
  manifests
    init.pp
    install.pp
    package.pp
    params.pp
  spec
  tests
  .fixtures.yml
  .rspec
  Gemfile
  Gemfile.lock
  metadata.json
  Rakefile
  README.md
```

We will add the following code to the file:

```puppet
class consul::package {

  package { 'bind':
    ensure => present
  } ->

  file { '/etc/named.conf':
    ensure  => present,
    content => template("consul/named.conf.erb"),
    mode    => '0644',
    owner   => 'root',
    group   => 'root',
    require => Package['bind'],
  } ~>

  file { '/etc/named/consul.conf':
    ensure  => present,
    content => template("consul/consul.conf.erb"),
    mode    => '0644',
    owner   => 'root',
    group   => 'root',
    require => Package['bind'],
  } ~>

  service { 'named':
    ensure  => running,
    enable  => true,
    require => File['/etc/named.conf'],
  }
}
```

Now, let's create a `templates` folder. In this example, we are not parameterizing the files, and in a production instance, you would. That's why we are using the `templates` folder and not files:

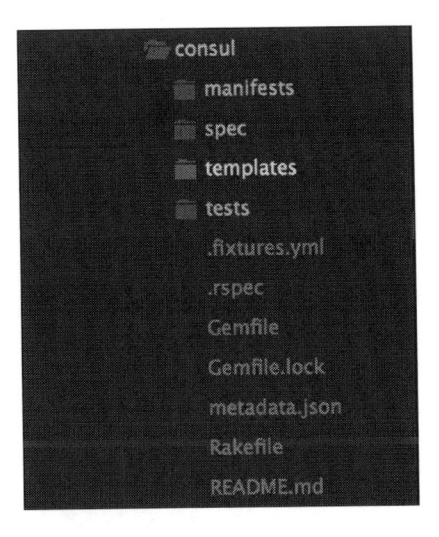

```
consul
  manifests
  spec
  templates
  tests
  .fixtures.yml
  .rspec
  Gemfile
  Gemfile.lock
  metadata.json
  Rakefile
  README.md
```

Now, let's create a file called `named.conf.erb` and add the following code to it:

```
options {
  listen-on port 53 { 127.0.0.1; };
  listen-on-v6 port 53 { ::1; };
  directory       "/var/named";
  dump-file       "/var/named/data/cache_dump.db";
  statistics-file "/var/named/data/named_stats.txt";
  memstatistics-file "/var/named/data/named_mem_stats.txt";
  allow-query     { localhost; };
  recursion yes;

  dnssec-enable no;
  dnssec-validation no;

  /* Path to ISC DLV key */
  bindkeys-file "/etc/named.iscdlv.key";

  managed-keys-directory "/var/named/dynamic";
};

include "/etc/named/consul.conf";
```

Configuring Service Discovery and Docker Networking

The code is just setting our DNS resolver to listen to `127.0.0.1`. Remember that we have set port forwarding on our Consul container to forward port `53`. That is how the host will connect to the container. Lastly, it will call our next template file `/etc/named/consul.conf`. Let's create that now:

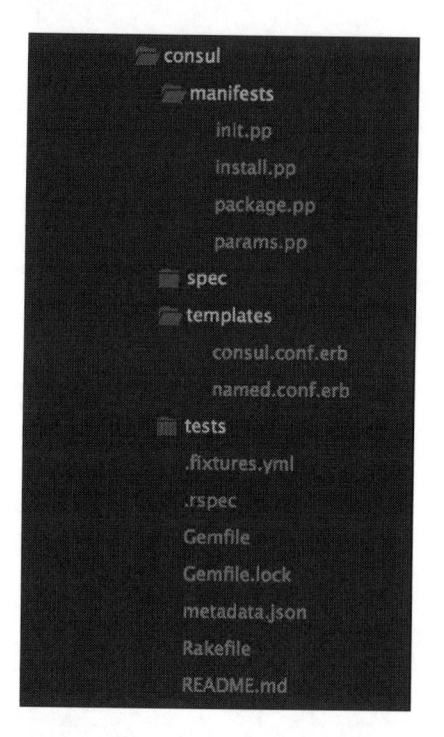

```
consul
    manifests
        init.pp
        install.pp
        package.pp
        params.pp
    spec
    templates
        consul.conf.erb
        named.conf.erb
    tests
    .fixtures.yml
    .rspec
    Gemfile
    Gemfile.lock
    metadata.json
    Rakefile
    README.md
```

The code that we will add is as follows:

```
zone "consul" IN {
    type forward;
    forward only;
    forwarders { 127.0.0.1 port 8600; };
};
```

You will note that we are forwarding port `8600`, which is the port that Consul uses for its DNS traffic, and removing port `53`. As TCP bind will use port `53`, we will forward the request to `8600`, as shown in the following piece of code:

```puppet
class consul::install {

  package { 'device-mapper-libs':
    ensure => '1.02.107-5.el7_2.1',
  }

  class { 'docker':
    version     => $consul::docker_version,
    tcp_bind    => $consul::docker_tcp_bind,
    socket_bind => 'unix:///var/run/docker.sock',
    require     => Package['device-mapper-libs']
  } ->

  docker::image { $consul::consul_docker_image : } ->

  docker::run { $consul::consul_container_hostname:
    image    => $consul::consul_docker_image,
    hostname => $consul::consul_container_hostname,
    command  => "-server --advertise ${consul::consul_advertise} -bootstrap-expect ${consul::consul_bootstrap_expect}",
    ports    => ['8301:8301', '8301:8301/udp', '8302:8302', '8302:8302/udp', '8400:8400', '8500:8500', '8600:8600', '8600:8600/udp']
  }

  docker::image { $consul::reg_docker_image : } ->

  docker::run { $consul::reg_container_hostname:
    image   => $consul::reg_docker_image,
    net     => $consul::reg_net,
    volumes => $consul::reg_volume,
    command => $consul::reg_command,
  }
}
```

We need to make one more change before we can run Puppet. We need to add the new code of `package.pp` to our `init.pp` file. We can do so like this:

```puppet
class consul (
  $docker_version           = $consul::params::docker_version,
  $docker_tcp_bind          = $consul::params::docker_tcp_bind,
  $consul_docker_image      = $consul::params::consul_docker_image,
  $consul_container_hostname = $consul::params::consul_container_hostname,
  $consul_advertise         = $consul::params::consul_advertise,
  $consul_bootstrap_expect  = $consul::params::consul_bootstrap_expect,
  $reg_docker_image         = $consul::params::reg_docker_image,
  $reg_container_hostname   = $consul::params::reg_container_hostname,
  $reg_net                  = $consul::params::reg_net,
  $reg_volume               = $consul::params::reg_volume,
  $reg_command              = $consul::params::reg_command,

) inherits consul::params {

  include consul::install
  include consul::package

}
```

Configuring Service Discovery and Docker Networking

Now, we can run our module. Let's go to the terminal and change to root of our Vagrant repo. We will issue the `vagrant up` command and if you already have a box running, just issue the `vagrant destroy -f && vagrant up` command. Now, let's check the web UI (`127.0.0.1:8500`):

As you can see in the preceding screenshot, we have a new service registered on port `8600` (`consul-8600`). Now, we need to make sure that our machines are listening to the right DNS servers on their interfaces. We are going to do this in `servers.yaml`, as I would usually add this configuration to my user data in AWS. You could very well control this with Puppet. So, in future, you can decide the right place for the configuration of your environment. The line we are going to add is - `{ shell: 'echo -e "PEERDNS=no\nDNS1=127.0.0.1\nDNS2=8.8.8.8">>/etc/sysconfig/network-scripts/ifcfg-enp0s3 && systemctl restart network'}`. We will add it as shown in the following screenshot:

[90]

Now, let's go to our terminal and issue the `vagrant up` command. If you have a box already running then issue the `vagrant destroy -f && vagrant up` command. The terminal output should look like the one shown in the following screenshot:

We can then log in to our vagrant box using `vagrant ssh` and test whether our DNS setup works. We can do this by selecting a service and trying to ping it. We are going to choose our `ping bitbucket-server-7990` service by entering the `ping bitbucket-server-7990.service.consul` command, and we should get the following results:

Configuring Service Discovery and Docker Networking

As you can see in the preceding screenshot, it returns the echo response as the loopback, as the service is running locally on this host. If we were external of the host, it would return the IP of the host that is running the service. Now, we run our container schedulers, such as Docker swarm, that have multiple host. We now know how service discovery works.

Now, let's have a look at what this would look like using Docker Compose.

In order to not repeat ourselves, let's make our `init.pp` file the same as the module that uses the `manifests` method. We have to make one small change to the `params.pp` file; `docker-compose` expects that you pass it strings. So, we need to remove the brackets around `$reg_volume` as shown in the following screenshot:

```
{  1   class consul::params {
   2
   3     $docker_version             = '1.9.1-1.el7.centos'
   4     $docker_tcp_bind            = 'tcp://127.0.0.1:4243'
   5     $consul_docker_image        = 'scottyc/consul'
   6     $consul_container_hostname  = 'consul'
   7     $consul_advertise           = $::ipaddress_enp0s8
   8     $consul_bootstrap_expect    = '1'
   9     $reg_docker_image           = 'gliderlabs/registrator'
  10     $reg_container_hostname     = 'registrator'
  11     $reg_net                    = 'host'
  12     $reg_volume                 = '/var/run/docker.sock:/tmp/docker.sock'
  13     $reg_command                = "consul://$::ipaddress_enp0s8:8500"
} 14   }
  15
```

Then, we will add our `package.pp` file as we did earlier and also create the two templates for our `bind` config. Then, we need to update our `docker-compose.yml.erb` file in our `templates` directory. We need to add our second container, `regisrator`. We are going to use the same parameters as we did in the manifest module earlier in this chapter. The code for this should look as shown in the following screenshot:

```
 1
 2  <%= @consul_container_hostname %>:
 3      image: <%= @consul_docker_image %>
 4      hostname: <%= @consul_container_hostname %>
 5      ports:
 6          - "8300:8300"
 7          - "8301:8301"
 8          - "8301:8301/udp"
 9          - "8302:8302"
10          - "8302:8302/udp"
11          - "8400:8400"
12          - "8500:8500"
13          - "8600:8600"
14          - "8600:8600/udp"
15      command: -server --advertise <%= @consul_advertise %> -bootstrap-expect <%= @consul_bootstrap_expect %>
16
17  registrator:
18      image: <%= @reg_docker_image %>
19      net: "<%= @reg_net %>"
20      volumes:
21          - <%= @reg_volume %>
22      command: "<%= @reg_command %>"
23
24
```

You will also note that we changed the ports on our Consul container as we did earlier in the chapter (we removed port 53 and added 8600 tcp/udp). Now, we can go to our terminal, change to root of our Vagrant repo, and issue the vagrant up command. Our terminal should look like the one shown in the following screenshot:

Configuring Service Discovery and Docker Networking

Again, we can also check our browser at `127.0.0.1:8500`:

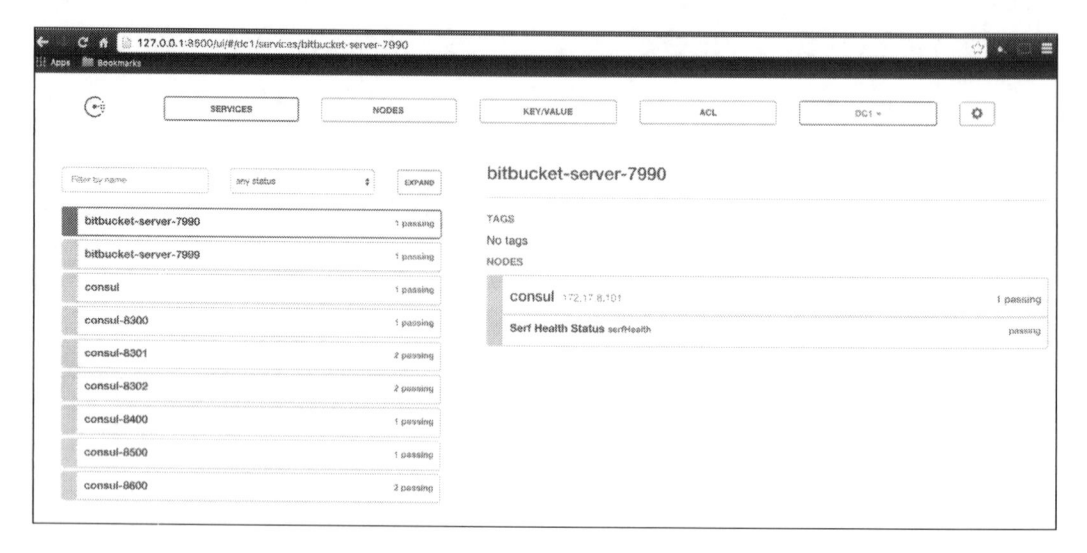

As you can see in the preceding screenshot, it looks the same as it did earlier in the chapter.

Let's log in to our box and test our DNS service discovery. For this, enter the `vagrant ssh` command and then ping a service. This time, we will choose something different. We will use the `ping consul-8500.service.consul` command. We should get the following response after this:

```
[vagrant@node-01 ~]$ ping consul-8500.service.consul
PING consul-8500.service.consul (127.0.0.1) 56(84) bytes of data.
64 bytes from node-01 (127.0.0.1): icmp_seq=1 ttl=64 time=0.040 ms
64 bytes from node-01 (127.0.0.1): icmp_seq=2 ttl=64 time=0.069 ms
64 bytes from node-01 (127.0.0.1): icmp_seq=3 ttl=64 time=0.123 ms
64 bytes from node-01 (127.0.0.1): icmp_seq=4 ttl=64 time=0.051 ms
^C
--- consul-8500.service.consul ping statistics ---
4 packets transmitted, 4 received, 0% packet loss, time 3000ms
rtt min/avg/max/mdev = 0.040/0.070/0.123/0.033 ms
[vagrant@node-01 ~]$
```

So that's all for service discovery in this chapter. We will be picking it up again in the container scheduler chapter.

Docker networking

In this topic, we are going to look at the native networking stack that comes with Docker Engine. There is a wealth of knowledge that you can achieve by reading on this subject. I strongly suggest that you do, as there is a lot you can do with Docker networking. If you have not used it before, I would suggest that you start reading the guide at https://docs.docker.com/engine/userguide/networking/dockernetworks/. From here, you can read about the different types of drivers, how to use VXLAN to separate your networks, and the best practices when designing your Docker network. We are going to cover the basics now and the more advanced features in later chapters.

The prerequisites

Before we can even start to code for our network, there are a few things we need. First, we need a key/value store. Docker will use this to map all the containers, IP addresses, and vxlans that are created. Seeing as there usually would be more than one host attached to a network, the key/value store is usually distributed to give it resiliency against failure. Luckily enough, we have already built a key/value store that we can take advantage of, it's Consul of course. The other configuration that you will need are extra args when we start our Docker Engine. This is to let Docker Engine know how to access the key/value store. These are the basic prerequisites that we need to get coding.

The code

Let's create our first Docker network. To do this, we are going to add to our `consul` module. I am not going to do this twice for both manifests and `docker-compose`, as the configuration can be ported between the two. I am going to use the `docker-compose` module for my example. If this is the first time that you are creating a Docker network, it would be a worth while exercise to port the configuration to both. So, lets' start. We are only going to make changes to our `install.pp` file. The first change that we are going to make is to our extra arguments for our `docker-engine` daemon. We do this by adding the code shown in the following screenshot:

```
class { 'docker':
    version       => $consul::docker_version,
    tcp_bind      => $consul::docker_tcp_bind,
    socket_bind   => 'unix:///var/run/docker.sock',
    extra_parameters => '--cluster-store=consul://127.0.0.1:8500 --cluster-advertise=enp0s8:2376',
    require       => Package['device-mapper-libs']
} ->
```

The code sets our key/value store's address and port. Then, it also tells other machines what interface and port we are advertising our networks on.

The next code we are going to add will create the network. We will create a new file called `network.pp`. Then, we will add the code shown in the following screenshot to it:

```puppet
class consul::network {

  docker_network { 'docker-internal':
    ensure => present,
    create => true,
    driver => 'overlay',
  }
}
```

The next thing we will have to do is make sure that our classes get installed in the correct order, as the Docker network is dependent on Consul being there. If Consul is not there, our catalogue will fail. So, we need to use the `contain` functionality built into Puppet. We do this by adding the code shown in the following screenshot:

```puppet
class consul (

  $docker_version            = $consul::params::docker_version,
  $docker_tcp_bind           = $consul::params::docker_tcp_bind,
  $consul_docker_image       = $consul::params::consul_docker_image,
  $consul_container_hostname = $consul::params::consul_container_hostname,
  $consul_advertise          = $consul::params::consul_advertise,
  $consul_bootstrap_expect   = $consul::params::consul_bootstrap_expect,
  $reg_docker_image          = $consul::params::reg_docker_image,
  $reg_container_hostname    = $consul::params::reg_container_hostname,
  $reg_net                   = $consul::params::reg_net,
  $reg_volume                = $consul::params::reg_volume,
  $reg_command               = $consul::params::reg_command,

) inherits consul::params {

  contain consul::install
  contain consul::package
  contain consul::network

  Class['consul::install'] -> Class['consul::package'] -> Class['consul::network']

}
```

[96]

As you can see, we are just setting up a basic network. We could set things such as IP address range, gateway, and so on. If we do that, it would look like this:

```
45    docker_network { 'my-net':
46        ensure   => present,
47        create   => true,
48        driver   => 'overlay',
49        subnet   => '192.168.1.0/24',
50        gateway  => '192.168.1.1',
51        iprange  => ' 192.168.1.4/32'
52    }
```

Now that we have our code, let's go to our terminal and issue the `vagrant up` command from root of our Vagrant repo. Our terminal output should look like the one shown in the following screenshot:

Now, we can check to make sure that our network is there by logging in to our vagrant box (`vagrant ssh` from the root of our Vagrant repo). Once we log in to our box, we need to change to root (`sudo -i`) and then issue the `docker network ls` command. This will list the available networks on the box. The one we are looking for is `docker-internal` with the `overlay` driver:

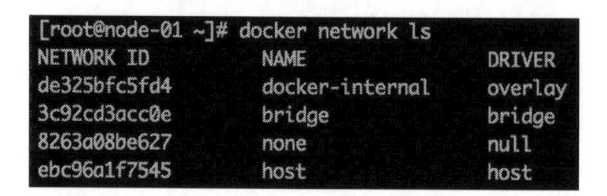

```
[root@node-01 ~]# docker network ls
NETWORK ID          NAME                DRIVER
de325bfc5fd4        docker-internal     overlay
3c92cd3acc0e        bridge              bridge
8263a08be627        none                null
ebc96a1f7545        host                host
```

As you can see from the output of our terminal, we were successful and our network has been configured. That is all we are going to do with networking in this chapter. In the next chapter, we will be attaching containers and spanning our Docker network across multiple hosts.

Summary

In this chapter, you learned a lot about how the container ecosystem handles service discovery. I can't emphasize on how important it will be to understand this topic when you start using containers at scale. I would really suggest that you get a solid understanding of service discovery before moving on to further chapters. We also covered the basics of Docker networking. Don't worry, as in the next chapter, we will go into Docker networking in more depth as we will be building multihost applications.

6
Multinode Applications

In this chapter, we are going to start playing with the really cool stuff. We are going to use all the skills we have learned in the book so far. We are really going to step it up by a notch. In this chapter, we are going to deploy four servers. We will look at how to cluster Consul, which will give us a perfect opportunity to further our modules' functionality. In this chapter, we are going to look at two ways to network our containers. First, by using the standard host IP network, that our Consul cluster will communicate on. We will also install the **ELK (Elasticsearch, Logstash, and Kibana)** stack (`https://www.elastic.co/`). To do this, we will be writing a module for each of the products. Because Elasticsearch is our data store in this solution, we want to hide it so that only Logstash and Kibana can access the application. We will accomplish this using the native Docker networking stack and isolate Elasticsearch using VXLAN. As you can see, we are going to get through a lot in this chapter. We will cover the following topics in this chapter:

- The design of our solution
- Putting it all together

The design of our solution

As there are a lot of moving parts in the solution, it would be best to visualize what we are going to be coding for. As this will be a big step up from our last chapter, we will break down the solution. In the first topic, we will look at the design for our Consul cluster.

Multinode Applications

The Consul cluster

In the design, we are going to use four servers: **node-01**, **node-02**, **node-03**, and **node-04**. We will use **node-01** to bootstrap our Consul cluster. We will add the other three nodes to the cluster as servers. They will be able to join the conciseness, vote, and replicate the key/value store. We will set up an IP network that is on the `172.17.8.0/24` network. We will map our container ports to the host ports that sit on the `172.17.8.0/24` network. The following image will show the network flow:

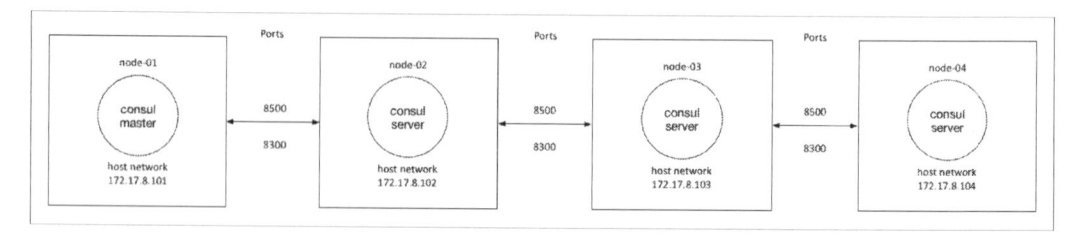

The ELK stack

Now that we have our Consul cluster, we can look at what our ELK stack is going to look like. First, we will walk through the design of the network. The stack will be connected to the native Docker network. Note that I have not listed the IP addresses of our Docker network. The reason for this is that we will let the Docker daemon select the networking address range. For this solution, we are not going to route any traffic out of this network, so letting the daemon choose the IP range is fine. So you will also note that Elasticsearch is only connected to our Docker network. This is because we only want Logstash and Kibana. This is to keep other applications from being able to send requests or queries to Elasticsearch. You will note that both Logstash and Kibana are connected to both the Docker network and the host network. The reason for this is that we want applications to send their logs to Logstash, and we will want to access the Kibana web application:

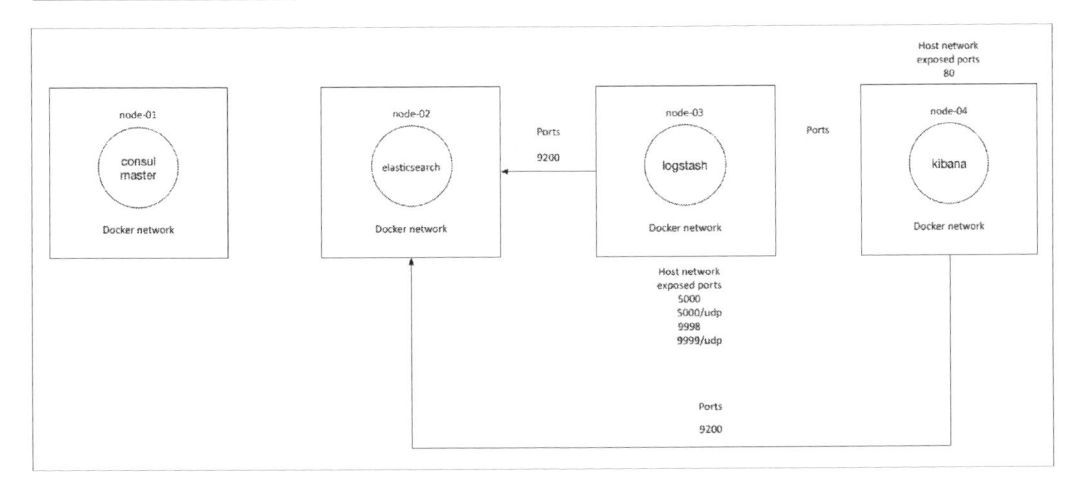

To get the full picture of the architecture, we just need to overlay both the diagrams. So, let's start coding!!!

Putting it all together

Now that we have looked at the design, let's put it all together. We will look at the changes to the plumbing of our Vagrant repo. Then, we will code the extra functionality into our `consul` module. We will then run our Vagrant repo to make sure that our Consul cluster is up and running. After we have completed that task, we will build the ELK stack, and we will build a module for each of the products. We will also set up Logstash to forward logs to node-03, just so we can test to make sure that our ELK stack is correct. Let's get some light on this.

Multinode Applications

The server setup

We will now look at the changes we are going to make to our new Vagrant repo. The first file that we are going to look at is our `servers.yaml` file. The first thing we need to do is change our base box. As we are going to be connecting containers to the native Docker network, our host machines must run a kernel version above 3.19. I have created a prebuilt vagrant box with just this configuration. It is the Puppetlabs box that we have been using in all the other chapters with the kernel updated to version 4.4:

```
---
-
  box: scottyc/centos-7-puppet-kernel-4-4
  cpu: 1
  ip: "172.17.8.101"
  name: node-01
  forward_ports:
    - { guest: 80, host: 8081 }
    - { guest: 8500, host: 8500 }
  ram: 2048
  shell_commands:
    - { shell: 'systemctl stop firewalld && systemctl disable firewalld' }
    - { shell: 'yum install -y wget git lvm2 device-mapper-libs' }
    - { shell: '/opt/puppet/bin/gem install r10k && ln -s /opt/puppet/bin/r10k /usr/bin/r10k || true' }
    - { shell: 'cp /home/vagrant/node-01/Puppetfile /tmp && cd /tmp && r10k puppetfile install --verbose' }
    - { shell: 'cp /home/vagrant/node-01/modules/* -R /tmp/modules || true' }
    - { shell: 'echo -e "172.17.8.101 node-01\n172.17.8.102 node-02\n172.17.8.103 node-03\n172.17.8.104 node-04">/etc/hosts' }
    - { shell: 'echo -e "PEERDNS=no\nDNS1=127.0.0.1\nDNS2=8.8.8.8">>/etc/sysconfig/network-scripts/ifcfg-enp0s3 && systemctl restart network' }
```

As you can note in the preceding screenshot, the other change that we have made to the `servers.yaml` file is that we have added entries to the `/etc/hosts` directory. We have done this to simulate a traditional DNS infrastructure. If this had been in your production environment, we wouldn't have needed to add that configuration.

Now, we have to add the other three servers. The following screenshot will show exactly how it should look:

```
box: scottyc/centos-7-puppet-kernel-4-4
cpu: 1
ip: "172.17.0.101"
name: node-01
forward_ports:
  - { guest: 80, host: 8061 }
  - { guest: 8500, host: 8500 }
ram: 2048
shell_commands:
  - { shell: 'systemctl stop firewalld && systemctl disable firewalld'}
  - { shell: 'yum install -y wget git lvm2 device-mapper-libs' }
  - { shell: '/opt/puppet/bin/gem install r10k && ln -s /opt/puppet/bin/r10k /usr/bin/r10k || true'}
  - { shell: 'cp /home/vagrant/node-01/Puppetfile /tmp && cd /tmp && r10k puppetfile install --verbose' }
  - { shell: 'cp /home/vagrant/node-01/modules/* -R /tmp/modules || true' }
  - { shell: 'echo -e "172.17.8.101 node-01\n172.17.8.102 node-02\n172.17.8.103 node-03\n172.17.8.104 node-04">/etc/hosts' }
  - { shell: 'echo -e "PEERDNS=no\nDNS1=127.0.0.1\nDNS2=8.8.8.8">>/etc/sysconfig/network-scripts/ifcfg-enp0s3 && systemctl restart network'}

box: scottyc/centos-7-puppet-kernel-4-4
cpu: 1
ip: "172.17.0.102"
name: node-02
forward_ports:
  - { guest: 80, host: 8082 }
ram: 2048
shell_commands:
  - { shell: 'systemctl stop firewalld && systemctl disable firewalld'}
  - { shell: 'yum install -y wget git lvm2 device-mapper-libs' }
  - { shell: '/opt/puppet/bin/gem install r10k && ln -s /opt/puppet/bin/r10k /usr/bin/r10k || true'}
  - { shell: 'cp /home/vagrant/node-02/Puppetfile /tmp && cd /tmp && r10k puppetfile install --verbose' }
  - { shell: 'cp /home/vagrant/node-02/modules/* -R /tmp/modules' }
  - { shell: 'echo -e "172.17.8.101 node-01\n172.17.8.102 node-02\n172.17.8.103 node-03\n172.17.8.104 node-04">/etc/hosts' }
  - { shell: 'echo -e "PEERDNS=no\nDNS1=127.0.0.1\nDNS2=8.8.8.8">>/etc/sysconfig/network-scripts/ifcfg-enp0s3 && systemctl restart network'}

box: scottyc/centos-7-puppet-kernel-4-4
cpu: 1
ip: "172.17.0.103"
name: node-03
forward_ports:
  - { guest: 80, host: 8083 }
ram: 2048
shell_commands:
  - { shell: 'systemctl stop firewalld && systemctl disable firewalld'}
  - { shell: 'yum install -y wget git lvm2 device-mapper-libs' }
  - { shell: '/opt/puppet/bin/gem install r10k && ln -s /opt/puppet/bin/r10k /usr/bin/r10k || true'}
  - { shell: 'cp /home/vagrant/node-03/Puppetfile /tmp && cd /tmp && r10k puppetfile install --verbose' }
  - { shell: 'cp /home/vagrant/node-03/modules/* -R /tmp/modules' }
  - { shell: 'echo -e "172.17.8.101 node-01\n172.17.8.102 node-02\n172.17.8.103 node-03\n172.17.8.104 node-04">/etc/hosts' }
  - { shell: 'echo -e "PEERDNS=no\nDNS1=127.0.0.1\nDNS2=8.8.8.8">>/etc/sysconfig/network-scripts/ifcfg-enp0s3 && systemctl restart network'}

box: scottyc/centos-7-puppet-kernel-4-4
cpu: 1
ip: "172.17.0.104"
name: node-04
forward_ports:
  - { guest: 80, host: 8080 }
ram: 2048
shell_commands:
  - { shell: 'systemctl stop firewalld && systemctl disable firewalld'}
  - { shell: 'yum install -y wget git lvm2 device-mapper-libs' }
  - { shell: '/opt/puppet/bin/gem install r10k && ln -s /opt/puppet/bin/r10k /usr/bin/r10k || true'}
  - { shell: 'cp /home/vagrant/node-04/Puppetfile /tmp && cd /tmp && r10k puppetfile install --verbose' }
  - { shell: 'cp /home/vagrant/node-04/modules/* -R /tmp/modules' }
  - { shell: 'echo -e "172.17.8.101 node-01\n172.17.8.102 node-02\n172.17.8.103 node-03\n172.17.8.104 node-04">/etc/hosts' }
  - { shell: 'echo -e "PEERDNS=no\nDNS1=127.0.0.1\nDNS2=8.8.8.8">>/etc/sysconfig/network-scripts/ifcfg-enp0s3 && systemctl restart network'}
```

So, the ports that we will hit once all our servers are built are `8500` on `node-01` (the Consul web UI `127.0.0.1:8500`) and `8081` (the Kibana web UI `127.0.0.1:8081`).

The Consul cluster

We are well aquatinted with the `consul` module now, but we are going to take it to the next level. For this chapter, we are just going to use the compose version. The reason why is because when you start getting into more complex applications or need to use an `if` statement to add logic, an `.erb` file gives us that freedom. There are a fair few changes to this module. So, let's start again at our `params.pp` file:

```puppet
class consul::params {

    $docker_version              = '1.9.1-1.el7.centos'
    $docker_tcp_bind             = 'tcp://127.0.0.1:4243'
    $consul_docker_image         = 'scottyc/consul'
    $consul_container_hostname   = "$hostname"
    $consul_advertise            = $::ipaddress_enp0s8
    $consul_bootstrap_expect     = '1'
    $consul_master_ip            = '172.17.8.101'
    $reg_docker_image            = 'gliderlabs/registrator'
    $reg_container_hostname      = 'registrator'
    $reg_net                     = 'host'
    $reg_volume                  = '/var/run/docker.sock:/tmp/docker.sock'
    $reg_command                 = "consul://$::ipaddress_enp0s8:8500"

    if ($::hostname == 'node-01') { $consul_is_master = true }
    else { $consul_is_master = false }
}
```

As you can see, we have added two new parameters. The first one is `$consul_master_ip` and the other parameter is `$consul_is_master`. We will use this to define which server will bootstrap our Consul cluster and which server will join the cluster. We have hardcoded the hostname of **node-01**. If this was a production module, I would not hardcode a hostname that should be a parameter that is looked up in Hiera (https://docs.puppetlabs.com/hiera/3.0/). We will pick up on this again when we look at our `docker-compose.yml.erb` file. The other parameters should look very familiar to you.

Next, let's look at our `init.pp` file:

```puppet
class consul {

    $docker_version            = $consul::params::docker_version,
    $docker_tcp_bind           = $consul::params::docker_tcp_bind,
    $consul_docker_image       = $consul::params::consul_docker_image,
    $consul_container_hostname = $consul::params::consul_container_hostname,
    $consul_advertise          = $consul::params::consul_advertise,
    $consul_bootstrap_expect   = $consul::params::consul_bootstrap_expect,
    $consul_master_ip          = $consul::params::consul_master_ip,
    $consul_is_master          = $consul::params::consul_is_master,
    $reg_docker_image          = $consul::params::reg_docker_image,
    $reg_container_hostname    = $consul::params::reg_container_hostname,
    $reg_net                   = $consul::params::reg_net,
    $reg_volume                = $consul::params::reg_volume,
    $reg_command               = $consul::params::reg_command,

    ) inherits consul::params {

    validate_bool($consul_is_master)

    contain consul::install
    contain consul::package
    contain consul::network

    Class['consul::install'] -> Class['consul::package'] -> Class['consul::network']

}
```

As you can see here, we have not changed this file much, as we have added a Boolean (`$consul_is_master`). However, we will want to validate the input. We do this by calling the stdlib function, `validate_bool`.

Let's quickly browse through the `install.pp` file:

```puppet
class consul::install {

    package { 'device-mapper-libs':
        ensure => installed,
    }

    class { 'docker':
        version          => $consul::docker_version,
        tcp_bind         => $consul::docker_tcp_bind,
        socket_bind      => 'unix:///var/run/docker.sock',
        extra_parameters => "--cluster-store=consul://$consul::consul_master_ip:8500 --cluster-advertise=enp0s8:2376",
        require          => Package['device-mapper-libs']
    } ->

    file { '/root/docker-compose.yml':
        ensure  => file,
        content => template('consul/docker-compose.yml.erb'),
    } ->

    docker_compose { $consul::container_hostname :
        ensure => present,
        source => '/root',
        scale  => ['1', '1']
    }
}
```

Multinode Applications

Now, let's look at the `network.pp` file:

```puppet
class consul::network {

    docker_network { 'docker-internal':
        ensure  => present,
        create  => true,
        driver  => 'overlay',
    }
}
```

Finally, we will look at the `package.pp` file:

```puppet
class consul::package {

    package { 'bind':
        ensure => present
    } ->

    file { '/etc/named.conf':
        ensure  => present,
        content => template("consul/named.conf.erb"),
        mode    => '0644',
        owner   => 'root',
        group   => 'root',
        require => Package['bind'],
    } ~>

    file { '/etc/named/consul.conf':
        ensure  => present,
        content => template("consul/consul.conf.erb"),
        mode    => '0644',
        owner   => 'root',
        group   => 'root',
        require => Package['bind'],
    } ~>

    service { 'named':
        ensure  => running,
        enable  => true,
        require => File['/etc/named.conf'],
    }
}
```

As you can see, we have not made any changes to these files. Now, we can look at the file that will really have the logic for deploying our container. We will then move to our `templates` folder and look at our `docker-compose.yml.erb` file. This is where most of the changes in the module have been made.

Chapter 6

So, let's look at the contents of the file, which are shown in the following screenshot:

```erb
<% if @consul_is_master == true then -%>
<%= @consul_container_hostname %>:
    image: <%= @consul_docker_image %>
    hostname: <%= @consul_container_hostname %>
    restart: always
    ports:
        - "8300:8300"
        - "8301:8301"
        - "8301:8301/udp"
        - "8302:8302"
        - "8302:8302/udp"
        - "8400:8400"
        - "8500:8500"
        - "8600:8600"
        - "8600:8600/udp"
    command: -server --client 0.0.0.0 --advertise <%= @consul_advertise %> -bootstrap-expect <%= @consul_bootstrap_expect %>
<% elsif @consul_is_master == false then -%>
<%= @consul_container_hostname %>:
    image: <%= @consul_docker_image %>
    hostname: <%= @consul_container_hostname %>
    ports:
        - "8300:8300"
        - "8301:8301"
        - "8301:8301/udp"
        - "8302:8302"
        - "8302:8302/udp"
        - "8400:8400"
        - "8500:8500"
        - "8600:8600"
        - "8600:8600/udp"
    command: -server -bind 0.0.0.0 --client 0.0.0.0  --advertise <%= @consul_advertise %> -join <%= @consul_master_ip %>
<% end -%>

registrator:
    image: <%= @reg_docker_image %>
    restart: always
    net: "<%= @reg_net %>"
    volumes:
        - <%= @reg_volume %>
    command: "<%= @reg_command %>"
```

So as you can see, the code in this file has doubled. Let's break it down into three pieces, as shown in the following screenshot:

```erb
<% if @consul_is_master == true then -%>
<%= @consul_container_hostname %>:
    image: <%= @consul_docker_image %>
    hostname: <%= @consul_container_hostname %>
    restart: always
    ports:
        - "8300:8300"
        - "8301:8301"
        - "8301:8301/udp"
        - "8302:8302"
        - "8302:8302/udp"
        - "8400:8400"
        - "8500:8500"
        - "8600:8600"
        - "8600:8600/udp"
    command: -server --client 0.0.0.0 --advertise <%= @consul_advertise %> -bootstrap-expect <%= @consul_bootstrap_expect %>
```

Multinode Applications

In the first block of code, the first change that you will note is the `if` statement. This is a choice to determine whether the node will be a master for bootstrapping Consul or a server in the cluster. If you remember from our `params.pp` file, we set `node-01` as our master. When we apply this class to our node, if its `node-01`, it will bootstrap the cluster. The next line that we want to pay attention to is as follows:

```
command: -server --client 0.0.0.0 --advertise <%= @consul_advertise
%> -bootstrap-expect <%= @consul_bootstrap_expect %>
```

We should just take note to compare the same line in the next block of code:

```
<% elsif @consul_is_master == false then -%>
<%= @consul_container_hostname %>:
    image: <%= @consul_docker_image %>
    hostname: <%= @consul_container_hostname %>
    ports:
      - "8300:8300"
      - "8301:8301"
      - "8301:8301/udp"
      - "8302:8302"
      - "8302:8302/udp"
      - "8400:8400"
      - "8500:8500"
      - "8600:8600"
      - "8600:8600/udp"
    command: -server -bind 0.0.0.0 --client 0.0.0.0  --advertise <%= @consul_advertise %> -join <%= @consul_master_ip %>
<% end -%>
```

First, we can see that this is `elsif`, the second half of our `if` statement. So, this will be the block of code that will install Consul on the other three nodes. They will still be servers in the cluster. They will just not have the job of bootstrapping the cluster. We can tell this from the following line:

```
command: -server -bind 0.0.0.0 --client 0.0.0.0  --advertise <%= @consul_advertise %> -join <%= @consul_master_ip %>
```

Remember earlier that we looked at the same line from the first line of code. You see the difference? In block one, we declare `-bootstrap-expect <%= @consul_bootstrap_expect %>`, and in the second block, we declare `-join <%= @consul_master_ip %>`. By looking at the code, this is how we can tell the bootstrap order. Lastly, we can see that we are declaring `<% end -%>` to close the `if` statement.

Now, let's look at the last block of code:

```
registrator:
    image: <%= @reg_docker_image %>
    restart: always
    net: "<%= @reg_net %>"
    volumes:
      - <%= @reg_volume %>
    command: "<%= @reg_command %>"
```

As you can see, it's going to deploy the `registrator` container. As this sits outside the `if` statement, this container will be deployed on any node that the `consul` class is applied to. We have made a lot of progress till now. We should check the module changes before moving on to creating our new elastic modules. The one last thing we need to change is our `default.pp` manifest file, which is as follows:

```
node 'node-01' {

}

node 'node-02' {
  include consul
}

node 'node-03' {
  include consul
}

node 'node-04' {
  include consul
}
```

As you can see, we have a node declaration for each node and have applied the `consul` class. Now, let's open our terminal change directory to the root of our Vagrant repo and issue the `vagrant up` command. This time, it will download a new base box from Hashicloud. So, depending on your Internet connection, this could take some time. Remember that the reason we need this new box is that it has an updated kernel to take advantage of the native Docker networking. In the last chapter, we were able to create a network, but we weren't able to connect containers to it. In this chapter, we will. Also, we are going to build four servers, so running Vagrant should take about 5 minutes. As soon as our first machine is up, we can log on to the consul web UI. There we can watch the progress as each node is joined.

As you can see in the following screenshot, our cluster is bootstrapped:

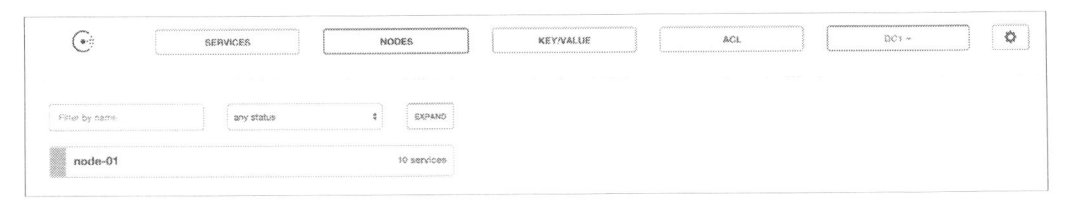

Multinode Applications

We can also check whether all our services are up and stable on the **SERVICES** tabs, as shown in this screenshot:

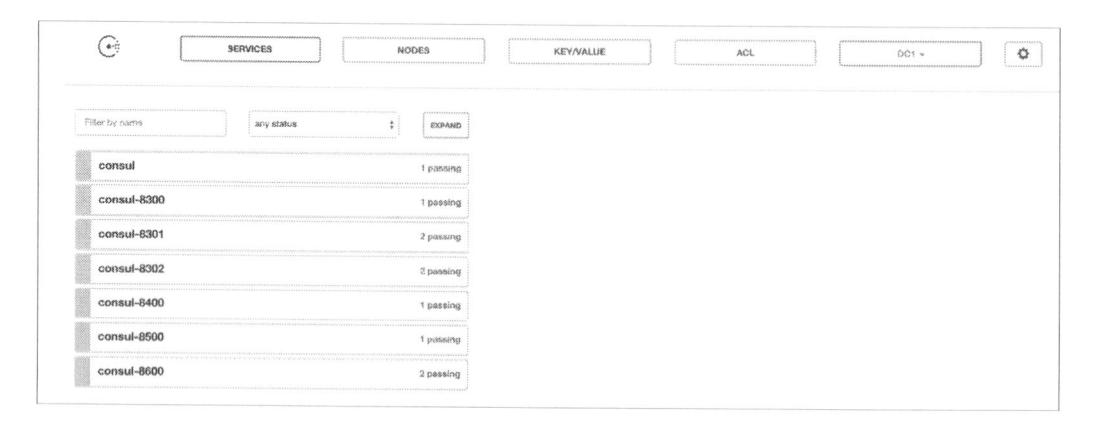

As you can see in the following screenshot, our second node has checked in:

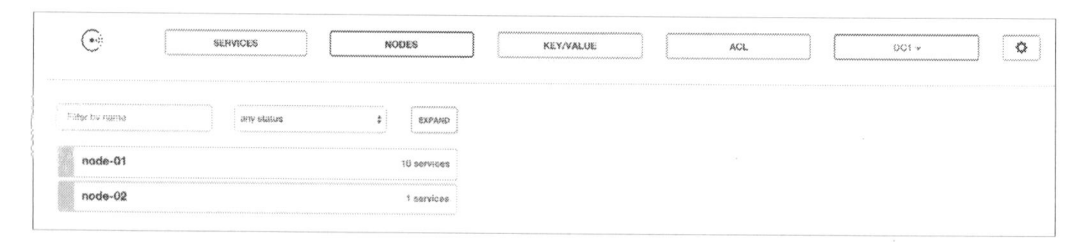

The following screenshot shows what the screen looks like when we go to our **SERVICES** tab:

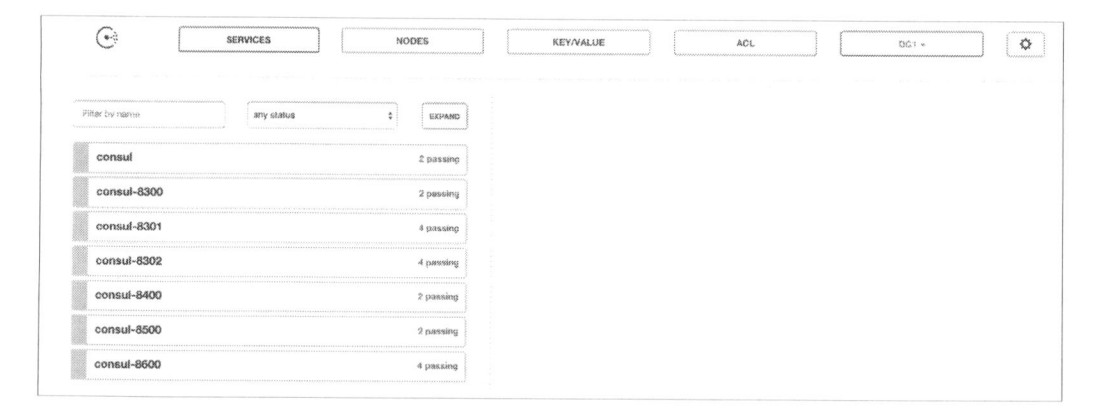

As you can see, our services have doubled. So things are looking good.

Now, the `vagrant up` command is complete, and our terminal output should look like the following screenshot:

Let's log back into our web browser to our Consul UI (`127.0.0.1:8500`). Under our **NODES** tab, we should now see all four nodes:

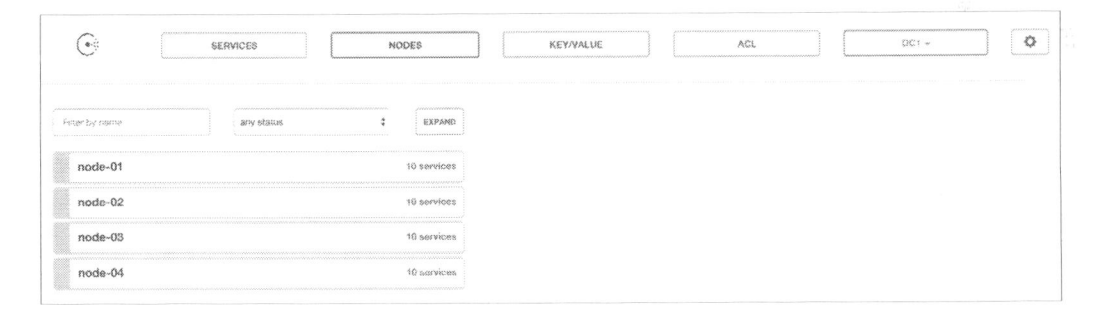

Multinode Applications

We can see that our cluster is in a good state as all four nodes have the same amount of services, which is **10**, and that all services are green. The last thing that we need to check is our DNS service discovery. So lets login to one of our boxes. We will choose **node-03**. So in our terminal, let's issue the `vagrant ssh node-03` command. We need to specify the node now as we have more than one vagrant box. We are going to ping the Consul service 8500. So, we just issue the `ping consul-8500.service.consul` command. The terminal output should look like the following screenshot:

```
[vagrant@node-03 ~]$ ping consul-8500.service.consul
PING consul-8500.service.consul (172.17.8.103) 56(84) bytes of data.
64 bytes from node-03 (172.17.8.103): icmp_seq=1 ttl=64 time=0.020 ms
64 bytes from node-03 (172.17.8.103): icmp_seq=2 ttl=64 time=0.054 ms
64 bytes from node-03 (172.17.8.103): icmp_seq=3 ttl=64 time=0.047 ms
64 bytes from node-03 (172.17.8.103): icmp_seq=4 ttl=64 time=0.047 ms
64 bytes from node-03 (172.17.8.103): icmp_seq=5 ttl=64 time=0.049 ms
64 bytes from node-03 (172.17.8.103): icmp_seq=6 ttl=64 time=0.048 ms
64 bytes from node-03 (172.17.8.103): icmp_seq=7 ttl=64 time=0.047 ms
64 bytes from node-03 (172.17.8.103): icmp_seq=8 ttl=64 time=0.048 ms
64 bytes from node-03 (172.17.8.103): icmp_seq=9 ttl=64 time=0.039 ms
64 bytes from node-03 (172.17.8.103): icmp_seq=10 ttl=64 time=0.114 ms
64 bytes from node-03 (172.17.8.103): icmp_seq=11 ttl=64 time=0.047 ms
64 bytes from node-03 (172.17.8.103): icmp_seq=12 ttl=64 time=0.046 ms
^C
--- consul-8500.service.consul ping statistics ---
12 packets transmitted, 12 received, 0% packet loss, time 11005ms
rtt min/avg/max/mdev = 0.020/0.050/0.114/0.021 ms
[vagrant@node-03 ~]$
```

This is now working perfectly. So, let's check one more thing now. We need to make sure that our Docker network is configured. For that, we will need to change the directory to root (`sudo -i`) and then issue the `docker network ls` command, as follows:

```
[root@node-03 ~]# docker network ls
NETWORK ID          NAME                DRIVER
b43a47a70589        docker-internal     overlay
ce2f0f870d53        none                null
290dbf24b3f6        host                host
2c0dc5e17f19        bridge              bridge
[root@node-03 ~]#
```

Now that everything is up and running, let's move on to our ELK stack.

The ELK stack

One of the focuses I had when planning this book was to use examples that could be ported to the real world so that readers get some real value. The ELK stack is no exception. The ELK stack is a very powerful stack of applications that lets you collate all your application logs to see the health of your application. For more reading on the ELK stack, visit `https://www.elastic.co/`. It has great documentation on all the products. Now, let's start our first new module.

As per our design, there is an order in which we need to install the ELK stack. Seeing as both Logstash and Kibana depend on Elasticsearch, we will build it first. All the images that we are going to use in our modules are built, maintained, and released by Elasticsearch. So we can be sure that the quality is good. The first thing that we need to do is create a new module called `<AUTHOR>-elasticsearch`. We covered how to create a module in the previous chapter, so if you are unsure how to do this, go back and read that chapter again. Now that we have our module, let's move it into the modules directory in the root of our Vagrant repo.

Seeing as the containers are already built by elastic, these modules are going to be short and sweet. We are just going to add code to the `init.pp` file:

```
class elasticsearch {
  docker::image { 'elasticsearch:2.1.0': } ->

  docker::run { 'elasticsearch':
    image     => 'elasticsearch:2.1.0',
    net       => 'docker-internal',
    command   => 'elasticsearch -Des.network.host=0.0.0.0',
    volumes   => ['/root/esdata:/usr/share/elasticsearch/data'],
    privileged => true,
  }
}
```

As you can see, we are calling the `docker::image` class to download `ealsticsearch`. In the `docker::run` class, we are calling our `elasticearch` container, and we are going to bind our container only to the `docker-internal` Docker network. You will note that we are not binding any ports. This is because, by default, this container will expose port 9200. We only want to expose port 9200 on the Docker network. Docker is smart enough to allow the exposed ports automatically on a Docker native network. In the next resource we are declaring just the host network for `elasticsearch`. We are specifying `0.0.0.0` as we don't know the IP that the container is going to get from the Docker network. As this service will be hidden from the outside world, this configuration will be fine. We will then map a persistent drive to keep our data.

Multinode Applications

The next thing that we need to do now is to add `elasticsearch` to a node. As per our design, we will add `elasticsearch` to `node-02`. We do this in our `default.pp` file in our `manifests` directory, as shown in the following screenshot:

```
node 'node-02' {
    include consul
    contain elasticsearch
}
```

You will note that I have used `contain` instead of `include`. This is because I want to make sure that the `consul` class is applied before the `elasticsearch` class, as we need our Docker network to be there before `elasticsearch` comes up. If the network is not there, our catalogue will fail as the container will not build.

The next module we are going to write is `logstash`. Our `logstash` module is going to be a little more complicated, as it will be on both the Docker network and the host network. The reason we want it on both is because we want applications to forward their logs to `logstash`. We also need `logstash` to talk to `elasticsearch`. Thus, we add `logstash` to the Docker network as well. We will create the module in the same way we did for `elasticsearch`. We will call our `<AUTHOR>-logstash` module. So, let's look at our code in our `init.pp` file, which is as follows:

```
class logstash {

    file { '/root/logstash-config':
        ensure => directory,
    } ->

    file { '/root/logstash-config/logstash.conf':
        ensure  => file,
        content => template("logstash/logstash.conf.erb"),
    } ->

    docker::image { 'logstash:2.1.0': } ->

    docker::run { 'logstash':
        image   => 'logstash:2.1.0',
        net     => 'docker-internal',
        volumes => ['/root/logstash-config:/opt/logstash/conf.d/'],
        ports   => ['9998:9998', '9999:9999/udp', '5000:5000', '5000:5000/udp'],
        env     => ['ES_HOST=elasticsearch', 'ES_PORT=9200'],
        command => 'logstash -f /opt/logstash/conf.d/logstash.conf --debug',
    }
}
```

Here, the first thing you will note is that we are creating a directory. This is to map to the container and will contain our `logstash.conf` file. The next declaration is a file type. This is our `logstah.conf` file, and as we can see, it's a template from the code. So, let's come back to it after looking at the rest of the code in our `init.pp` file. The next line of code will pull our `logstash` image from Docker Hub. In the `docker::run` class we will call our `logstash` container, use the `logstash` image, and attach the container to our `docker-internal` Docker network.

The next line of code will tell `logstash` to start using our `logstash.conf` file; we will then mount the directory we created earlier in our `init.pp` file to the container. Now, you can see in this module that we've exposed ports to the host network. In the last line, we tell `logstash` about our Elasticsearch host and Elasticsearch port. How does Logstash know where Elasticsearch is? We are not linking the containers like we did in previous chapters. This works in the same way to when we named our Elasticsearch container `elasticsearch`, and our Docker network has an inbuilt DNS server that lives at the address `127.0.0.11`. Any container that joins that network will register itself as its container name. This is how services on the `docker-internal` network find each other.

The last thing we need to look at is our template file for our `logstash.conf` file that we declared in our `init.pp` file. So, create a new folder called `templates` in the root of our module and then a file called `logstash.conf.erb`. We will add the following configuration to accept logs from syslog and Docker.

Multinode Applications

Lastly, at the bottom, we put our Elasticsearch configuration, as shown in this screenshot:

```
input {
  tcp {
    port => 5000
    type => syslog
  }
  udp {
    port => 5000
    type => syslog
  }
  file {
    type => "syslog"
    path => [ "/var/log/*.log", "/var/log/messages", "/var/log/syslog" ]
    start_position => "beginning"
  }
  file {
    type => "logstash"
    path => [ "/var/log/logstash/logstash.log" ]
    start_position => "beginning"
  }
}
filter {
  if [type] == "docker" {
    json {
      source => "message"
    }
    mutate {
      rename => [ "log", "message" ]
    }
    date {
      match => [ "time", "ISO8601" ]
    }
  }
  if [type] == "syslog" {
    grok {
      match => { "message" => "%{SYSLOG5424PRI}%{NONNEGINT:ver} +(?:%{TIMESTAMP_ISO8601:ts}|-) +(?:%{HOSTNAME:containerid}|-) +(?:%{NOTSPACE:containername}|-) +(?:%{NOTSPACE:proc}|-) +(?:%{WORD:msgid}|-) +(?:%{SYSLOG5424SD:sd}|-) +(?:%{GREEDYDATA:msg})" }
    }
    syslog_pri { }
    date {
      match => [ "syslog_timestamp", "MMM d HH:mm:ss", "MMM dd HH:mm:ss" ]
    }
    if !("_grokparsefailure" in [tags]) {
      mutate {
        replace => [ "@source_host", "%{syslog_hostname}" ]
        replace => [ "@message", "%{syslog_message}" ]
      }
    }
    mutate {
      remove_field => [ "syslog_hostname", "syslog_message", "syslog_timestamp" ]
    }
  }
}
output {
  elasticsearch { hosts => ["elasticsearch:9200"] }
  stdout { codec => rubydebug }
}
```

Now, let's add our `logstash` module to `node-03` in the same way that we did with our `elastcsearch` module.

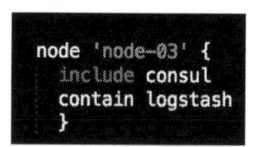

```
node 'node-03' {
  include consul
  contain logstash
}
```

Again, we will use `contain` instead of `include`. Now it's time to move on to our last module. We will create this in the same way as we have done for the last two modules. We will call this module `<AUTHOR>-kibana`.

In Kibana, we will only be adding code to the `init.pp` file, as shown in the following screenshot:

```
class kibana {

  docker::image { 'kibana:4.3.0': } ->

  docker::run { 'kibana':
    image => 'kibana:4.3.0',
    net   => 'docker-internal',
    ports => ['80:5601'],
    env   => ['ELASTICSEARCH_URL=http://elasticsearch:9200']
  }
}
```

As you can see, we are downloading the `kibana` image. In the `docker::run` class, we are calling our `kibana` container using the `kibana` image, attaching the container to our local Docker network. In the next line, we are mapping the container port `5601` (Kibana's default port) to port `80` on the host. This is just for ease of use for our lab. In the last line, we are telling `kibana` how to connect to `elasticsearch`.

Let's add `kibana` to `node-04` again using `contain` instead of `include`:

```
node 'node-01' {
    include consul
}

node 'node-02' {
    include consul
    contain elasticsearch
}

node 'node-03' {
    include consul
    contain logstash
}

node 'node-04' {
    include consul
    contain kibana
}
```

Multinode Applications

We are now ready to run our Vagrant environment. Let's open our terminal and change the directory to the root of our Vagrant repo. We will build this completely from scratch, so let's issue the `vagrant destroy -f && vagrant up` command.

This will take about 5 minutes or so to build, depending on your Internet connection, so please be patient. Once the build is complete, our terminal should have no errors and look like the following screenshot:

The next thing we will check is our Consul web UI (`127.0.0.1:8500`):

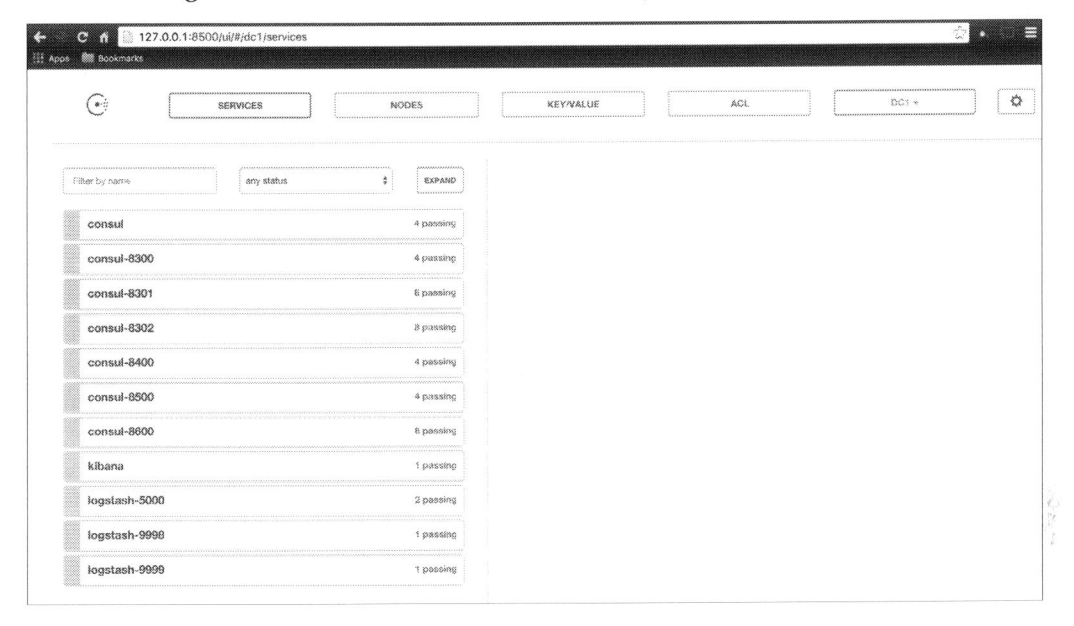

In the preceding screenshot, you can see that our Logstash and Kibana services are there, but where is Elasticsearch ? Don't worry, Elasticsearch is there, but we can't see it in Consul as we have not forwarded any ports to the host network. Registrator will only register services with exposed ports. We can make sure that our ELK stack is configured by logging in to our Kibana web UI (`127.0.0.1:8080`):

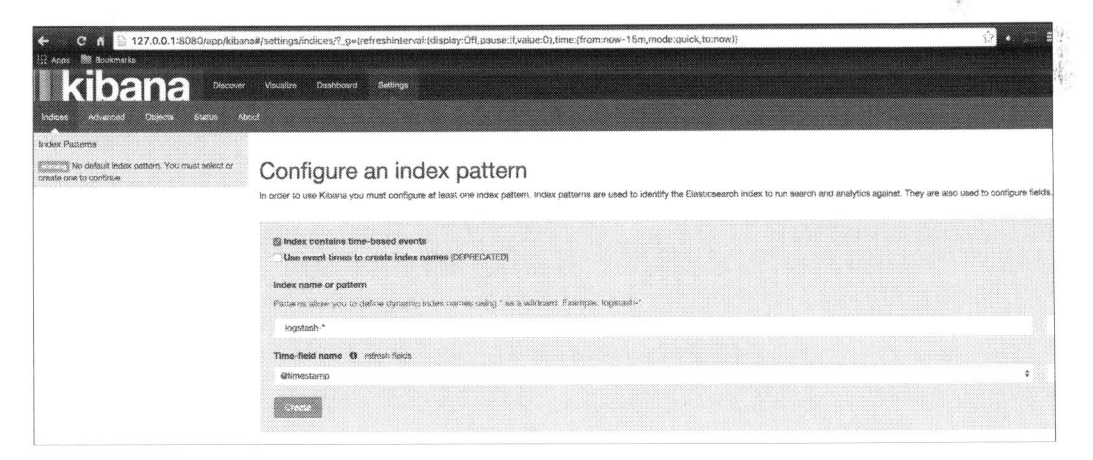

Multinode Applications

The next thing we need to do is click on the **Create** button. Then, if we go to the **Discover** tab, we can see the logs from Logstash:

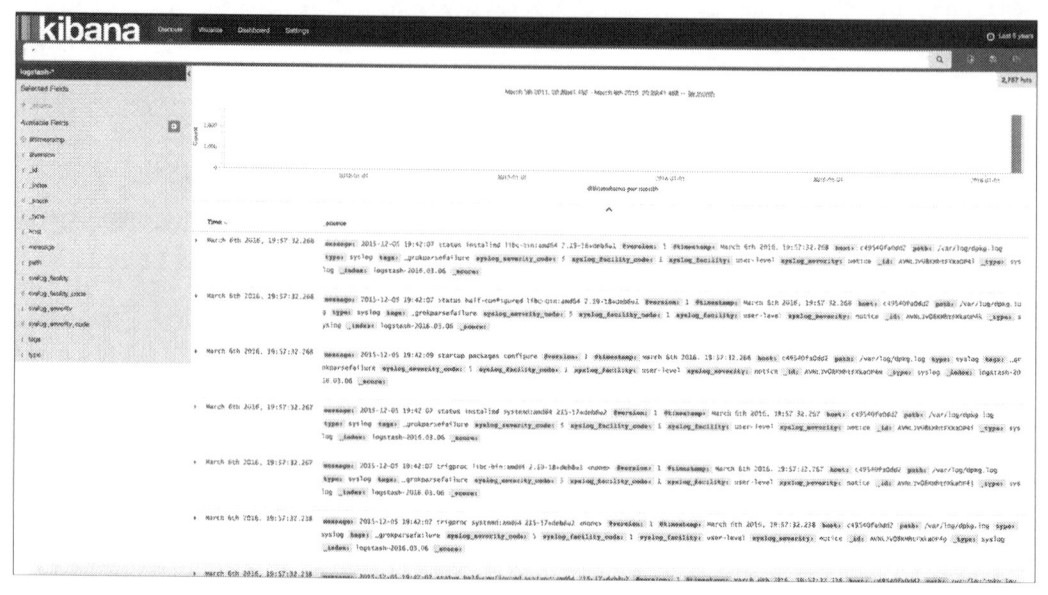

Logstash's logs

Summary

In this chapter, we looked at how to use Puppet to deploy containers across multiple nodes. We took advantage of the native Docker networking to hide services. This is a good security practice when working with production environments. The only issue with this chapter is that we don't have any failover or resiliency in our applications. This is why container schedulers are so important.

In the next chapter, we will drive into three different schedulers to arm you with the knowledge that you will need to make sound design choices in the future.

7
Container Schedulers

Now, we are at the business end of the book. There is a lot of buzz at the moment around this topic. This is where containers are going to go in future, and schedulers solve a lot of problems, for instance, spreading the load of our application across multiple hosts for us pending on load and starting our containers on another instance if the original host fails. In this chapter, we will look at three different schedulers. First, we will look at Docker Swarm. This is a Docker open source scheduler. We will build five servers and look at how to create a replicated master. We will then run a few containers and look at how Swarm will schedule them across nodes. The next scheduler we will look at is Docker **UCP (Universal Control Plane)**. This is a Docker enterprise solution that is integrated with Docker Compose. We will build a three-node cluster and deploy our Consul module. As UCP has a graphical interface, we will look at how UCP is scheduled from there. The final scheduler we will look at is Kubernetes. This is Google's offering and is also open source. For Kubernetes, we will build a single node using containers and use Puppet to define the more complex types. As you can see, we are going to look at each one differently, as they all have their individual strengths and weaknesses. Depending on your use case, you might decide on one or all of them to solve a problem that you may face.

Docker Swarm

For our first scheduler, we are going to look at Docker Swarm. This is a really solid product and in my opinion is a bit underrated compared to Kubernetes. It has really come on in leaps and bounds in the last few releases. It now supports replicated masters, rescheduling containers on failed hosts. So, let's look at the architecture of what we are building. Then, we will get into the coding.

The Docker Swarm architecture

In this example, we are going to build five servers, where two will be replicated masters and the other three nodes will be in the swarm cluster. As Docker Swarm needs a key/value store backend, we will use Consul. In this instance, we are not going to use our Consul modules; instead, we are going to use `https://forge.puppetlabs.com/KyleAnderson/consul`. The reason for this is that in all three examples, we are going to use different design choices. So, when you are trying to build a solution, you are exposed to more than one way to skin the cat. In this example, we are going to install Swarm and Consul onto the OS using Puppet and then run containers on top of it.

Coding

In this example, we are going create a new Vagrant repo. So, we will Git clone `https://github.com/scotty-c/vagrant-template.git` into the directory of our choice. The first thing that we will edit is the Puppetfile. This can be found in the root of our Vagrant repo. We will add the following changes to the file:

```ruby
#!/usr/bin/ruby env

require "socket"
$hostname = Socket.gethostname

forge 'http://forge.puppetlabs.com'


mod 'puppetlabs/stdlib'
mod 'puppetlabs/vcsrepo'
mod 'nanliu/staging'
mod 'KyleAnderson/consul'
mod 'scottyc/docker_swarm'
mod 'scottyc/golang'
mod 'garethr/docker', :git => "https://github.com/scotty-c/garethr-docker.git"
mod 'stankevich/python'
mod 'stahnma/epel'
mod 'maestrodev/wget'
```

The next file we will edit is `servers.yaml`. Again, this is located in the root of our Vagrant repo. We are going to add five servers to it. So, I will break down this file into five parts, one for each server.

First, let's look at the code for server 1:

```
box: scottyc/centos-7-puppet-kernel-4-4
cpu: 1
ip: "172.17.8.101"
name: swarm-101
forward_ports:
    - { guest: 8500, host: 9501 }
    - { guest: 80, host: 8001 }
    - { guest: 443, host: 8441 }
    - { guest: 8080, host: 8081 }
ram: 4096
shell_commands:
    - { shell: yum install -y git wget curl lvm2 unzip device-mapper-libs && systemctl stop firewalld && systemctl disable firewalld }
    - { shell: 'echo -e "PEERDNS=no\nDNS1=127.0.0.1\nDNS2=8.8.8.8">>/etc/sysconfig/network-scripts/ifcfg-enp0s3 && systemctl restart network'}
    - { shell: /opt/puppet/bin/gem install r10k }
    - { shell: 'echo -e "172.17.8.101 swarm-101">/etc/hosts && echo "PATH=\$PATH:/usr/local/bin" >> ~/.bashrc' }
    - { shell: cp /home/vagrant/swarm-101/Puppetfile /tmp && cd /tmp && /opt/puppet/bin/r10k puppetfile install -v }
    - { shell: cp /home/vagrant/swarm-101/modules/* -R /tmp/modules }
```

Now, let's look at the code for server 2:

```
box: scottyc/centos-7-puppet-kernel-4-4
cpu: 1
ip: "172.17.8.102"
name: swarm-102
forward_ports:
    - { guest: 8500, host: 9502 }
    - { guest: 80, host: 8002 }
    - { guest: 443, host: 8442 }
    - { guest: 8080, host: 8082 }
ram: 4096
shell_commands:
    - { shell: yum install -y git wget curl lvm2 device-mapper-libs unzip && systemctl stop firewalld && systemctl disable firewalld }
    - { shell: 'echo -e "PEERDNS=no\nDNS1=127.0.0.1\nDNS2=8.8.8.8">>/etc/sysconfig/network-scripts/ifcfg-enp0s3 && systemctl restart network'}
    - { shell: /opt/puppet/bin/gem install r10k }
    - { shell: 'echo -e "172.17.8.101 swarm-101\n172.17.8.102 swarm-102">/etc/hosts && echo "PATH=\$PATH:/usr/local/bin" >> ~/.bashrc' }
    - { shell: cp /home/vagrant/swarm-102/Puppetfile /tmp && cd /tmp && /opt/puppet/bin/r10k puppetfile install -v }
    - { shell: cp /home/vagrant/swarm-102/modules/* -R /tmp/modules }
```

The following screenshot shows the code for server 3:

```
box: scottyc/centos-7-puppet-kernel-4-4
cpu: 1
ip: "172.17.8.103"
name: swarm-103
forward_ports:
    - { guest: 8500, host: 9503 }
    - { guest: 80, host: 8003 }
    - { guest: 443, host: 8443 }
    - { guest: 8080, host: 8083 }
ram: 4096
shell_commands:
    - { shell: yum install -y git wget curl lvm2 unzip device-mapper-libs && systemctl stop firewalld && systemctl disable firewalld }
    - { shell: 'echo -e "PEERDNS=no\nDNS1=127.0.0.1\nDNS2=8.8.8.8">>/etc/sysconfig/network-scripts/ifcfg-enp0s3 && systemctl restart network'}
    - { shell: /opt/puppet/bin/gem install r10k }
    - { shell: 'echo -e "172.17.8.101 swarm-101\n172.17.8.103 swarm-103">/etc/hosts && echo "PATH=\$PATH:/usr/local/bin" >> ~/.bashrc' }
    - { shell: cp /home/vagrant/swarm-103/Puppetfile /tmp && cd /tmp && /opt/puppet/bin/r10k puppetfile install -v }
    - { shell: cp /home/vagrant/swarm-103/modules/* -R /tmp/modules }
```

Container Schedulers

The following screenshot shows the code for server 4:

```
box: scottyc/centos-7-puppet-kernel-4-4
cpu: 1
ip: "172.17.8.114"
name: swarm-master-01
forward_ports:
    - { guest: 8500, host: 9504 }
ram: 2048
shell_commands:
    - { shell: yum install -y git wget curl lvm2 unzip device-mapper-libs && systemctl stop firewalld && systemctl disable firewalld }
    - { shell: 'echo -e "PEERDNS=no\nDNS1=127.0.0.1\nDNS2=8.8.8.8">>/etc/sysconfig/network-scripts/ifcfg-enp0s3 && systemctl restart network'}
    - { shell: /opt/puppet/bin/gem install r10k }
    - { shell: 'echo -e "172.17.8.101 swarm-101\n172.17.8.114 swarm-master-01">/etc/hosts && echo "PATH=\$PATH:/usr/local/bin" >> ~/.bashrc' }
    - { shell: cp /home/vagrant/swarm-master-01/Puppetfile /tmp && cd /tmp && /opt/puppet/bin/r10k puppetfile install -v }
    - { shell: cp /home/vagrant/swarm-master-01/modules/* -R /tmp/modules }
```

Finally, the code for server 5 is as follows:

```
box: puppetlabs/centos-7.0-64-puppet-enterprise
cpu: 1
ip: "172.17.8.115"
name: swarm-master-02
forward_ports:
    - { guest: 8500, host: 9505 }
ram: 2048
shell_commands:
    - { shell: yum install -y git wget curl lvm2 unzip device-mapper-libs && systemctl stop firewalld && systemctl disable firewalld }
    - { shell: 'echo -e "PEERDNS=no\nDNS1=127.0.0.1\nDNS2=8.8.8.8">>/etc/sysconfig/network-scripts/ifcfg-enp0s3 && systemctl restart network'}
    - { shell: /opt/puppet/bin/gem install r10k }
    - { shell: 'echo -e "172.17.8.101 swarm-101\n172.17.8.115 swarm-master-02">/etc/hosts && echo "PATH=\$PATH:/usr/local/bin" >> ~/.bashrc' }
    - { shell: cp /home/vagrant/swarm-master-02/Puppetfile /tmp && cd /tmp && /opt/puppet/bin/r10k puppetfile install -v }
    - { shell: cp /home/vagrant/swarm-master-02/modules/* -R /tmp/modules }
```

All this should look fairly familiar to you. The one call out is that we have used servers one through three as our cluster nodes. Four and five will be our masters.

The next thing we are going to do is add some values to Hiera. This is the first time we have used Hiera, so let's look at our `hiera.yaml` file to see our configuration in the root of our Vagrant repo, which is as follows:

```
1  :backends:
2    - yaml
3  :hierarchy:
4    - global
5
6  :yaml:
7    :datadir: /vagrant/hieradata
```

As you can see, we have a basic hierarchy. We just have one file to look at, which is our `global.yaml` file. We can see that the file lives in our `hieradata` folder as it is declared as our `data` directory. So, if we open the `global.yaml` file, we are going to add the following values to it:

```
docker_swarm::swarm_version: v1.1.3
consul::version: 0.6.3
```

[124]

The first value will tell Swarm what version we want to use. The last value sets the version of Consul that we will be using, which is `0.6.3`.

The next thing we need to do is write our module that will deploy both our Consul and Swarm clusters. We have created quite a few modules so far in the book, so we won't cover it again. In this instance, we will create a module called `<AUTHOR>-config`. We will then move the module into our `modules` folder at the root of our Vagrant repo. Now, let's look at what we are going to add to our `init.pp` file:

```puppet
# == Class: config
#
#
class config(

  $consul_ip = "$::ipaddress_enp0s8",

){

  include config::consul_config
  contain config::swarm
  contain config::compose
  contain config::dns
  contain config::run_containers

  Class['config::swarm'] -> Class['config::compose'] -> Class['config::dns'] -> Class['config::run_containers']
}
```

As you can see, this sets up our module. We will need to create a `.pp` file for each of our entries in our `init.pp` file, for example, `consul_config.pp`, `swarm.pp`, and so on. We have also declared one variable called `consul_ip`, whose value will actually come from Hiera, as that is where we set the value earlier. We will look at our files in alphabetical order. So, we will start with `config::compose.pp`, which is as follows:

```puppet
class config::compose {

  if $hostname =~ /^swarm-master*/ {

    notice ["This server is the Swarm Manager."]

  }

  else {

    file { '/root/docker-compose.yml':
      ensure  => file,
      content => template("config/registrator.yml.erb"),
    } ->

    docker_compose {'swarm app':
      ensure => present,
      source => '/root',
      scale  => ['1']
    }
  }
}
```

Container Schedulers

In this class, we are setting up a registrator. We are going to use Docker Compose for this, as it is a familiar configuration and something that we have already covered. You will note that there is an `if` statement in the code. This is a logic to run the container only on a cluster member. We don't want to run container applications on our masters. The next file we will look at is `consul_config.pp`, which is as follows:

```
class config::consul_config {

  if $hostname =~ /^*101*$/ {

    class { 'consul':
      config_hash => {
        'datacenter'       => 'dev',
        'data_dir'         => '/opt/consul',
        'ui_dir'           => '/opt/consul/ui',
        'bind_addr'        => $::ipaddress_enp0s8,
        'client_addr'      => '0.0.0.0',
        'node_name'        => "$::hostname",
        'advertise_addr'   => '172.17.8.101',
        'bootstrap_expect' => '1',
        'server'           => true
      }
    }
  }
  else {

    class { 'consul':
      config_hash => {
        'bootstrap'      => false,
        'datacenter'     => 'dev',
        'data_dir'       => '/opt/consul',
        'ui_dir'         => '/opt/consul/ui',
        'bind_addr'      => $::ipaddress_enp0s8,
        'client_addr'    => '0.0.0.0',
        'node_name'      => "$::hostname",
        'advertise_addr' => $::ipaddress_enp0s8,
        'start_join'     => ['172.17.8.101','172.17.8.103','172.17.8.103'],
        'server'         => false
      }
    }
  }

  ::consul::service { 'docker-service':
    checks  => [
      {
        script   => 'service docker status',
        interval => '10s',
        tags     => ['docker-service']
      }
    ],
    address => $::ipaddress_enp0s8,
  }
}
```

In this class, we will configure Consul, something that we have already covered in this book. I always like to look at multiple ways to do the same job, as you never know the solution that you might need to produce tomorrow and you always want the right tool for the right job. So, in this instance, we will not configure Consul in a container but natively on the host OS.

You can see that the code is broken into three blocks. The first block bootstraps our Consul cluster. You will note that the configurations are familiar, as they are the same as those we used to set up our container in an earlier chapter. The second block of code sets the members of the cluster after the cluster is bootstrapped. We have not seen this third block before. This sets up a service for Consul to monitor. This is just a taste of what we can manage, but we will look at this in anger in the next chapter. Let's look at our next file, that is, dns.pp:

```
class config::dns {

  package { 'bind':
    ensure => present
  } ->

  file { '/etc/named.conf':
    ensure  => present,
    content => template("config/named.conf.erb"),
    mode    => '0644',
    owner   => 'root',
    group   => 'root',
    require => Package['bind'],
  } ~>

  file { '/etc/named/consul.conf':
    ensure  => present,
    content => template("config/consul.conf.erb"),
    mode    => '0644',
    owner   => 'root',
    group   => 'root',
    require => Package['bind'],
  } ~>

  service { 'named':
    ensure  => running,
    enable  => true,
    require => File['/etc/named.conf'],
  }
}
```

Container Schedulers

You will note that this file is exactly the same as we used in our `consul` module. So, we can move on to the next file, which is `run_containers.pp`:

```puppet
class config::run_containers {

  if $hostname =~ /swarm-master-02/ {

    swarm_run {'jenkins':
      ensure  => present,
      image   => 'jenkins',
      ports   => ['8080:8080'],
      require => Class['config::swarm']
    }

    swarm_run {'nginx':
      ensure     => present,
      image      => 'nginx',
      ports      => ['80:80', '443:443'],
      log_driver => 'syslog',
      network    => 'swarm-private',
      require    => Class['config::swarm']
    }

    swarm_run {'redis':
      ensure  => present,
      image   => 'redis',
      network => 'swarm-private',
      require => Class['config::swarm']
    }
  }
}
```

This is the class that will run our containers on our cluster. At the top, we are declaring that we want master number two to initiate the call to the cluster. We are going to deploy three container applications. The first is `jenkins`, and as you can see, we will expose Jenkins web port `8080` to the host that the container runs on. The next container is `nginx`, and we are going to forward both `80` and `443` to the host, but also connect `nginx` to our private network across our `swarm-private` cluster. To get our logs from `nginx`, we will tell our container to use syslog as the logging driver. The last application that we will run is `redis`. This will be the backend of `nginx`. You will note that we are not forwarding any ports, as we are hiding Redis on our internal network. Now that we have our containers sorted, we have one file left, `swarm.pp`. This will configure our Swarm cluster and internal Swarm network as follows:

```puppet
class config::swarm {

  class { 'docker_swarm':
    require => Class['config::consul_config']
  }

  docker_network { 'swarm-private':
    ensure  => present,
    create  => true,
    driver  => 'overlay',
    require => Class['config::consul_config']
  }

  if $hostname =~ /^swarm-master*/ {

    swarm_cluster {'cluster 1':
      ensure       => present,
      backend      => 'consul',
      cluster_type => 'manage',
      port         => '8500',
      address      => '172.17.8.101',
      advertise    => $::ipaddress_enp0s8,
      path         => 'swarm',
    }
  }

  else {

    swarm_cluster {'cluster 1':
      ensure       => present,
      backend      => 'consul',
      cluster_type => 'join',
      port         => '8500',
      address      => '172.17.8.101',
      path         => 'swarm'
    }
  }
}
```

The first resource declaration will install the Swarm binaries. The next resource will configure our Docker network on each host. The next `if` state will define if the Swarm node is a master or part of the cluster via `hostname`. As you can see, we are declaring a few defaults that are the first values that tell Swarm what backend to use Consul with. The second value tells Swarm the IP address of the Consul backend which is `172.17.8.101`. The third value tells Swarm about the port it can access Swarm on, `8500`. The fourth value tells Swarm which interface to advertise the cluster on, which in our case is `enp0s8`. The last value sets the root of the key value store Swarm will use. Now, let's create our `templates` folder in the root of our module.

Container Schedulers

We will create three files there. The first file will be `consul.conf.erb` and the contents will be as follows:

```
zone "consul" IN {
    type forward;
    forward only;
    forwarders { 127.0.0.1 port 8600; };
};
```

The next file will be `named.conf.erb` and its contents will be as follows:

```
options {
  listen-on port 53 { 127.0.0.1; };
  listen-on-v6 port 53 { ::1; };
  directory       "/var/named";
  dump-file       "/var/named/data/cache_dump.db";
  statistics-file "/var/named/data/named_stats.txt";
  memstatistics-file "/var/named/data/named_mem_stats.txt";
  allow-query     { localhost; };
  recursion yes;

  dnssec-enable no;
  dnssec-validation no;

  /* Path to ISC DLV key */
  bindkeys-file "/etc/named.iscdlv.key";

  managed-keys-directory "/var/named/dynamic";
};

include "/etc/named/consul.conf";
```

The last file will be `registrator.yml.erb`, and the file's content will be as follows:

```
registrator:
  image: gliderlabs/registrator
  net: "host"
  volumes:
    - "/var/run/docker.sock:/tmp/docker.sock"
  command: consul://<%= @consul_ip %>:8500
```

The next thing that we need to add is our `config` class to our `default.pp` manifest file in the `manifest` folder in the root of our Vagrant repo:

```
1   node 'swarm-101' { include config }
2
3   node 'swarm-102' { include config }
4
5   node 'swarm-103' { include config }
6
7   node 'swarm-master-01' { include config }
8
9   node 'swarm-master-02' { include config }
10
```

Now, our module is complete and we are ready to run our cluster. So, let's open our terminal and change the directory to the root of our Vagrant repo and issue the `vagrant up` command. Now, we are building five servers, so be patient. Once our last server is built, our terminal output should look like that shown in this screenshot:

Terminal output

[131]

Container Schedulers

We can now look at our Consul cluster's web UI at `127.0.0.1:9501` (remember that we changed our ports on our `servers.yaml` file) as follows:

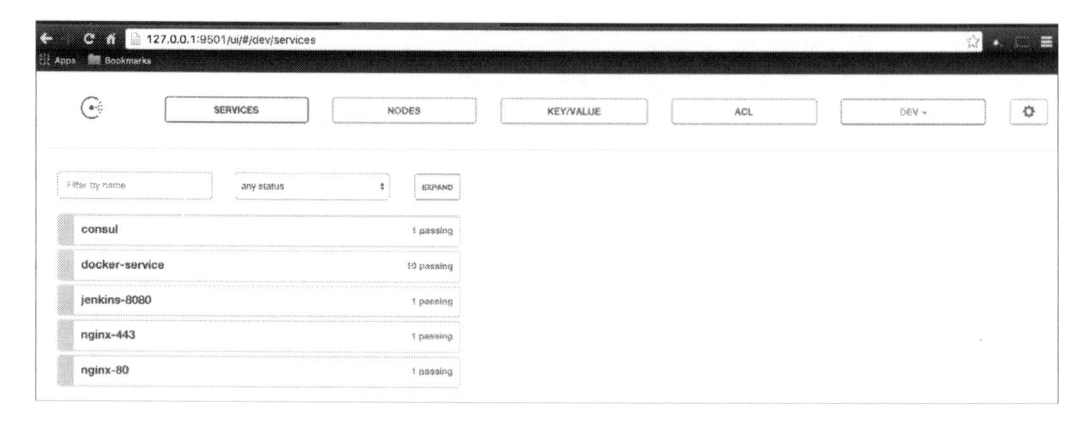

Now, let's see what host our Jenkins service is running on:

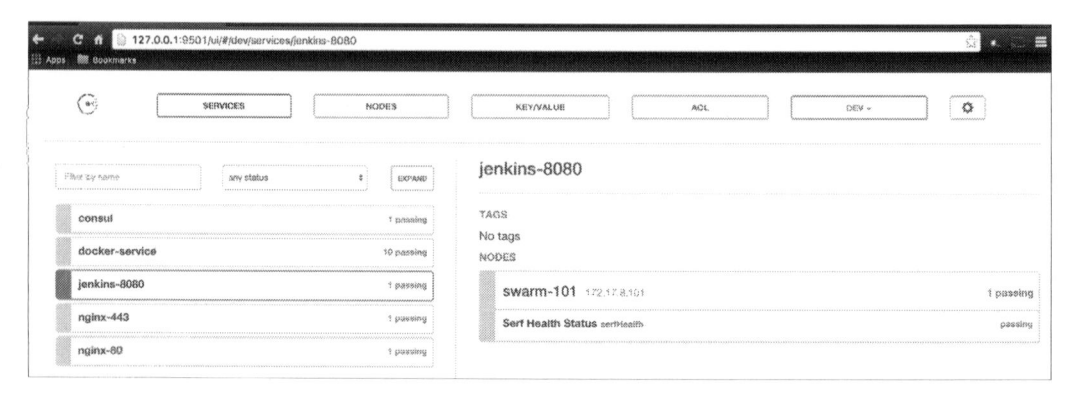

In this example, my service has come up on cluster node 101. You could get a different host for your cluster. So, we need to check what port `8080` forwards to our `servers.yaml` file guest. In my example, it's `8081`. So, if I go to my web browser and open a new tab and browse to `127.0.0.1:8081`, we will get the Jenkins page, which is as follows:

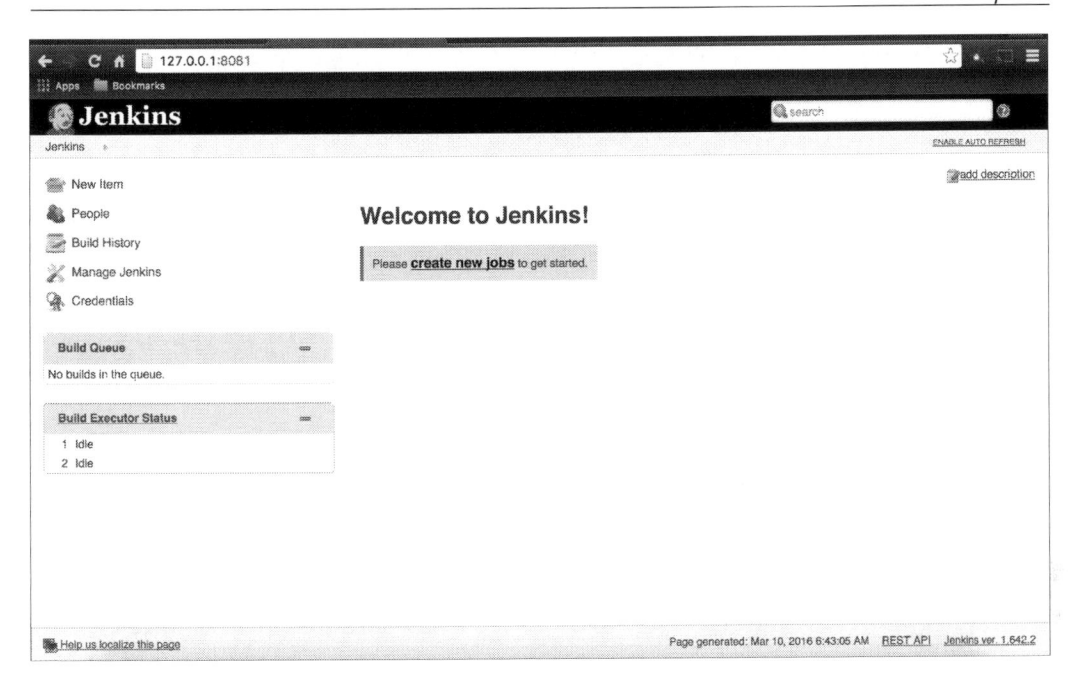

You can do the same with nginx, and I will leave that up to you as a challenge.

Docker UCP

In this topic, we will be looking at Docker's new offering, UCP. This product is not open sourced, so there is a licensing cost. You can get a trial license for 30 days (`https://www.docker.com/products/docker-universal-control-plane`), and that is what we will be using. Docker UCP takes out the complexity of managing all the moving parts of a scheduler. This could be a massive plus pending your use case. Docker UCP also brings with it a web UI for administration. So, if container schedulers seem daunting, this could be a perfect solution for you.

The Docker UCP architecture

So, in this example, we are going to build three nodes. The first will be our UCP controller and the other two nodes will be UCP HA replicas that give the design some fault tolerance. As UCP is a wrapped product, I am not going to go into all the moving parts.

Container Schedulers

Refer to the following diagram to get a visualization of the main components:

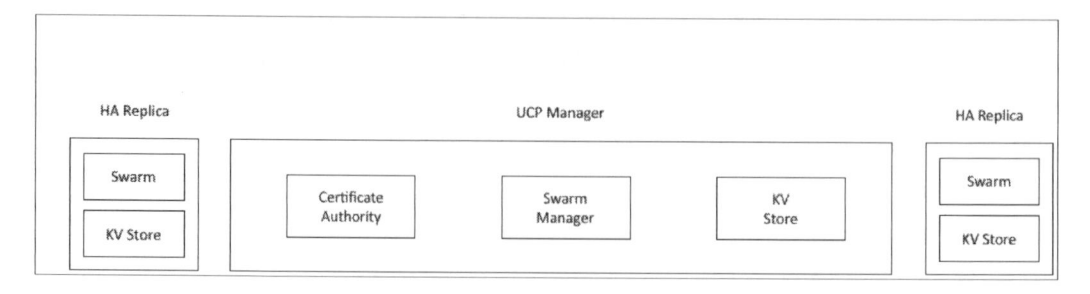

Coding

We are going to do something different in this module, just to show you that our modules are portable to any OS. We will build this module with the Ubuntu 14.04 server. Again, for this environment, we are going to create a new Vagrant repo. So, let's Git clone our Vagrant repo (`https://github.com/scotty-c/vagrant-template.git`). As in the last topic, we will look at the plumbing first, before we write our module. The first thing we are going to do is create a file called `config.json`. This will have your Docker Hub auth in the file:

```
{
    "auths": {
        "https://index.docker.io/v1/": {
            "auth": "xxxxxxxxxxxxxxxxxxxxxx",
            "email": "your.email@mail.com"
        }
    }
}
```

The next will be `docker_subscription.lic`. This will contain your trial license.

Now, let's look at our `servers.yaml` file in the root of our Vagrant repo, which is as follows:

```yaml
---
-
  box: puppetlabs/ubuntu-14.04-64-puppet-enterprise
  cpu: 2
  ip: "172.17.10.101"
  name: ucp-01
  forward_ports:
    - { guest: 443, host: 8443 }
  ram: 4096
  shell_commands:
    - { shell: 'apt-get update -y' }
    - { shell: 'apt-get install -y wget git' }
    - { shell: 'mkdir ~/.docker || true && cp /vagrant/config.json ~/.docker/' }
    - { shell: 'mkdir /etc/docker/ || true && cp /vagrant/docker_subscription.lic /etc/docker/subscription.lic' }
    - { shell: '/opt/puppet/bin/gem install r10k && ln -s /opt/puppet/bin/r10k /usr/bin/r10k || true'}
    - { shell: 'cp /home/vagrant/ucp-01/Puppetfile /tmp && cd /tmp && r10k puppetfile install --verbose' }
    - { shell: 'cp /home/vagrant/ucp-01/modules/* -R /tmp/modules || true' }

-
  box: puppetlabs/ubuntu-14.04-64-puppet-enterprise
  cpu: 2
  ip: "172.17.10.102"
  name: ucp-02
  forward_ports:
    - { guest: 443, host: 9443 }
  ram: 4096
  shell_commands:
    - { shell: 'apt-get update -y' }
    - { shell: 'apt-get install -y wget git' }
    - { shell: 'mkdir ~/.docker || true && cp /vagrant/config.json ~/.docker/' }
    - { shell: 'mkdir /etc/docker/ || true && cp /vagrant/docker_subscription.lic /etc/docker/subscription.lic' }
    - { shell: '/opt/puppet/bin/gem install r10k && ln -s /opt/puppet/bin/r10k /usr/bin/r10k || true'}
    - { shell: 'cp /home/vagrant/ucp-02/Puppetfile /tmp && cd /tmp && r10k puppetfile install --verbose' }
    - { shell: 'cp /home/vagrant/ucp-02/modules/* -R /tmp/modules || true' }

-
  box: puppetlabs/ubuntu-14.04-64-puppet-enterprise
  cpu: 2
  ip: "172.17.10.103"
  name: ucp-03
  forward_ports:
    - { guest: 443, host: 10443 }
  ram: 4096
  shell_commands:
    - { shell: 'apt-get update -y' }
    - { shell: 'apt-get install -y wget git' }
    - { shell: 'mkdir ~/.docker || true && cp /vagrant/config.json ~/.docker/' }
    - { shell: 'mkdir /etc/docker/ || true && cp /vagrant/docker_subscription.lic /etc/docker/subscription.lic' }
    - { shell: '/opt/puppet/bin/gem install r10k && ln -s /opt/puppet/bin/r10k /usr/bin/r10k || true'}
    - { shell: 'cp /home/vagrant/ucp-03/Puppetfile /tmp && cd /tmp && r10k puppetfile install --verbose' }
    - { shell: 'cp /home/vagrant/ucp-03/modules/* -R /tmp/modules || true' }
```

The main call out here is that now we are using the `puppetlabs/ubuntu-14.04-64-puppet-enterprise` vagrant box. We have changed `yum` to `apt-get`. Then, we are copying both our `config.json` and `docker_subscription.lic` files to their correct place on our vagrant box.

Container Schedulers

Now, we will look at the changes we need to make to our Puppetfile:

```ruby
#!/usr/bin/ruby env

require "socket"
$hostname = Socket.gethostname

forge 'http://forge.puppetlabs.com'

mod 'garethr/docker', :git => "https://github.com/scotty-c/garethr-docker.git"
mod 'puppetlabs/apt'
mod 'puppetlabs/docker_ucp'
mod 'puppetlabs/stdlib'
```

You will see that we need a few new modules from the Forge. The Docker module is familiar, as is stdlib. We will need Puppetlab's `apt` module to control our repos that Ubuntu will use to pull Docker. The last module is the Puppetlabs module for UCP itself. To find out more information about this module, you can read all about it at https://forge.puppetlabs.com/puppetlabs/docker_ucp. We will write a module that wraps this class and configures it for our environment.

Now, let's look at our Hiera file in `hieradata/global.yaml`:

```yaml
ucpconfig::ucp_url: https://172.17.10.101
ucpconfig::ucp_fingerprint:
```

As you can see, we have added two values. The first is `ucpconfig::ucp_url:`, which we will set to our first vagrant box. The next is the value for `ucpconfig::ucp_fingerprint:`, which we will leave blank for the moment. But remember it, as we will come back to this later in the topic.

Now, we will create a module called `<AUTHOR>-ucpconfig`. We have done this a few times now, so once you have created a module, create a folder in the root of our Vagrant repo called `modules` and move `ucpconfig` into that folder.

We will then create three manifests files in the module's `manifest` directory. The first will be `master.pp`, the second will be `node.pp`, and the last file will be `params.pp`.

Now, let's add our code to the `params.pp` file, as follows:

```
class ucpconfig::params {

    $ucp_url         = ''
    $ucp_username    = 'admin'
    $ucp_password    = 'orca'
    $ucp_fingerprint = ''
}
```

As you can see, we have four values: `Ucp_url`, which comes from Hiera; `ucp_username`, whose default value is set to `admin`; we then have `ucp_password` whose default value is set to `orca`. The last value is `ucp_fingerprint`, which again comes from Hiera. Now, in a production environment, I would set both the username and password in Hiera and overwrite the defaults, which we have set in `params.pp`. In this case, as it is a test lab, we will just use the defaults.

The next file we will look at is our `init.pp` file, which is as follows:

```
class ucpconfig {

    $ucp_url         = $ucpconfig::params::ucp_url,
    $ucp_username    = $ucpconfig::params::ucp_username,
    $ucp_password    = $ucpconfig::params::ucp_password,
    $ucp_fingerprint = $ucpconfig::params::ucp_fingerprint,
) inherits ucpconfig::params {

class { 'docker':
    socket_bind => 'unix:///var/run/docker.sock',
}

case $::hostname {
  'ucp-01': {
    include ucpconfig::master
  }
  default: {
    include ucpconfig::node
  }
}
}
```

You can see that at the top of the class, we are mapping our `params.pp` file. The next declaration installs the `docker` class and sets the `socket_bind` parameter for the daemon. Now, the next bit of logic defines whether the node is a master or a node depending on the hostname. As you can see, we are only setting `ucp-01` as our master.

Now, let's look at `master.pp`:

```
class ucpconfig::master(

    $ucp_url         = $ucpconfig::ucp_url,
    $ucp_username    = $ucpconfig::ucp_username,
    $ucp_password    = $ucpconfig::ucp_password,
    $ucp_fingerprint = $ucpconfig::ucp_fingerprint,

) {

    class { 'docker_ucp':
        controller                 => true,
        host_address               => $::ipaddress_eth1,
        version                    => '1.0.1',
        usage                      => false,
        tracking                   => false,
        subject_alternative_names  => $::ipaddress_eth0,
        external_ca                => false,
        swarm_scheduler            => 'binpack',
        swarm_port                 => 19001,
        controller_port            => 443,
        preserve_certs             => true,
        docker_socket_path         => '/var/run/docker.sock',
        license_file               => '/etc/docker/subscription.lic',
        require                    => Class['docker']
    }
}
```

In this class, we have the logic to install the UCP controller or master. At the top of the class, we are mapping our parameters to our `init.pp` file. The next block of code calls the `docker_ucp` class. As you can see, we are setting the value of controller to `true`, the host address to our second interface, the alternate name of the cluster to our first interface, and the version to `1.0.1` (which is the latest at the time of writing this book). We will then set the ports for both the controller and Swarm. Then, we will tell UCP about the Docker socket location and also the location of the license file.

Now, let's look at our last file, `node.pp`:

```puppet
class ucpconfig::node (

    $ucp_url         = $ucpconfig::ucp_url,
    $ucp_username    = $ucpconfig::ucp_username,
    $ucp_password    = $ucpconfig::ucp_password,
    $ucp_fingerprint = $ucpconfig::ucp_fingerprint,

){

    class { 'docker_ucp':
        ucp_url                   => $ucp_url,
        fingerprint               => $ucp_fingerprint,
        username                  => $ucp_username,
        password                  => $ucp_password ,
        host_address              => $::ipaddress_eth1,
        subject_alternative_names => $::ipaddress_eth0,
        replica                   => true,
        version                   => '1.0.1',
        usage                     => false,
        tracking                  => false,
        require                   => Class['docker']
    }
}
```

As you can see, most of the settings might look familiar. The call out for a node is that we need to point it to the controller URL (which we set in Hiera). We will get to know about the admin username and password and the cluster fingerprint just a little bit later. So, that completes our module. We now need to add our class to our nodes which we will do by adding the `default.pp` manifest file in the `manifests/default.pp` location from the root of our Vagrant repo, as follows:

```puppet
node 'ucp-01' {
    include ucpconfig
}


node 'ucp-02' {
    include ucpconfig
}


node 'ucp-03' {
    include ucpconfig
}
```

Container Schedulers

Let's go to our terminal and change the directory to the root of our Vagrant repo. This time, we are going to do something different. We are going to issue the `vagrant up ucp-01` command. This will bring up only the first node. We do this as we need to get the fingerprint that is generated as UCP comes up.

Our terminal output should look like that shown in the following screenshot:

Terminal output

You will note that the fingerprint has been displayed on your terminal output. For my example, the fingerprint is `INFO[0031] UCP Server SSL: SHA1 Fingerprint=C2:7C:BB:C8:CF:26:59:0F:DB:BB:11:BC:02:18:C4:A4:18:C4:05:4E`. So, we will add this to our Hiera file, which is `global.yaml`:

```
ucpconfig::ucp_url: https://172.17.10.101
ucpconfig::ucp_fingerprint: C2:7C:BB:C8:CF:26:59:0F:DB:BB:11:BC:02:18:C4:A4:18:C4:05:4E
```

Now that we have our first node up, we should be able to log in to the web UI. We do this in our browser. We will go to `https:127.0.0.1:8443` and get the login page as follows:

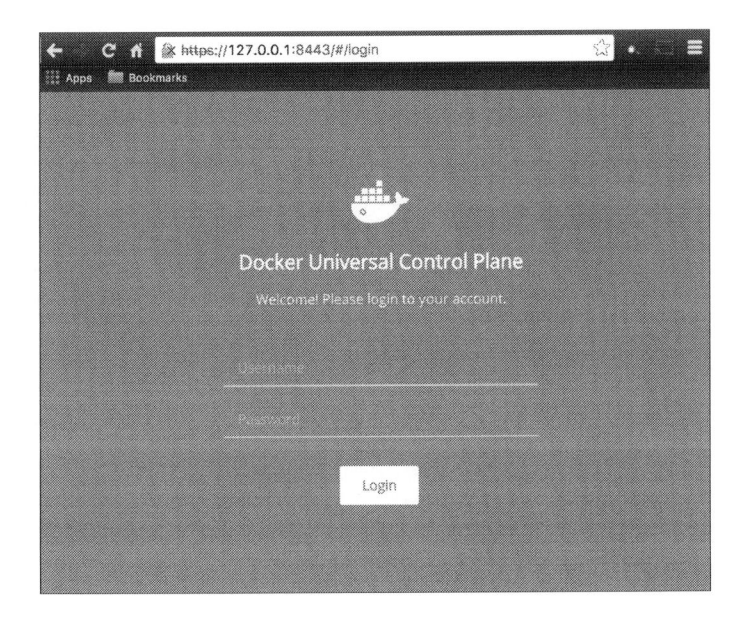

We will then add the username and password that we set in our `params.pp` file:

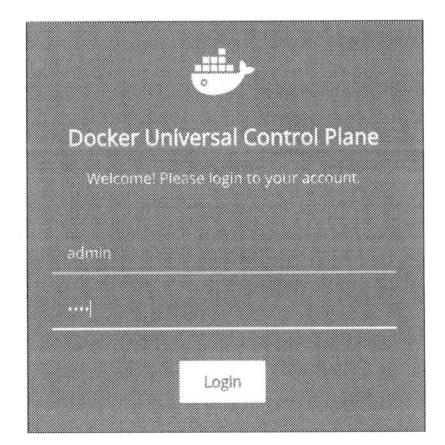

Container Schedulers

Then, after we log in, you can see that we have a health cluster, as follows:

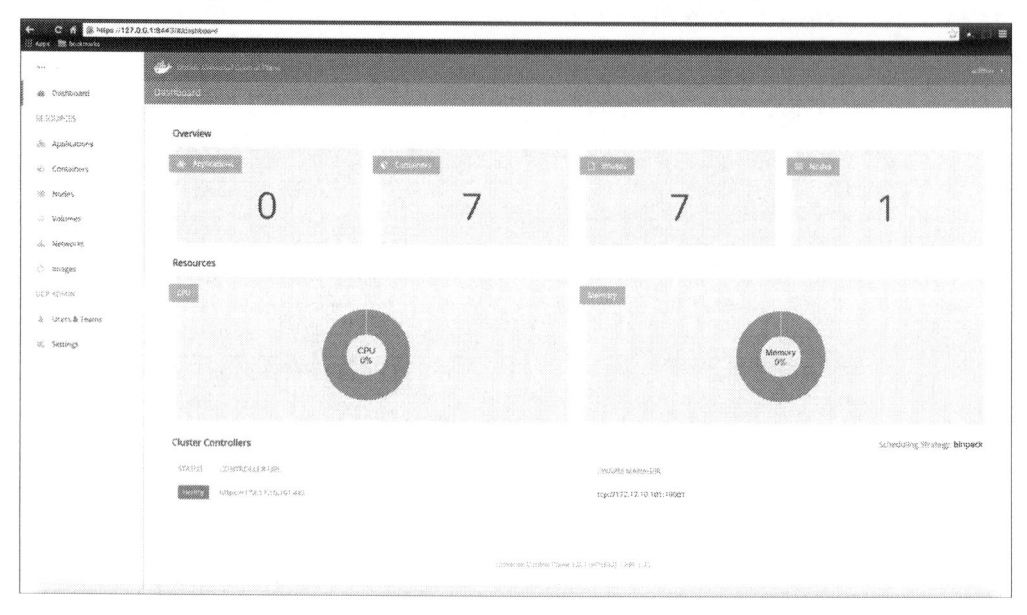

Health cluster after logging in

Now, let's return to our terminal and issue the `vagrant up ucp-02 && vagrant up ucp-03` command.

Once that is complete, if we look at our web UI, we can see that we have three nodes in our cluster, which are as follows:

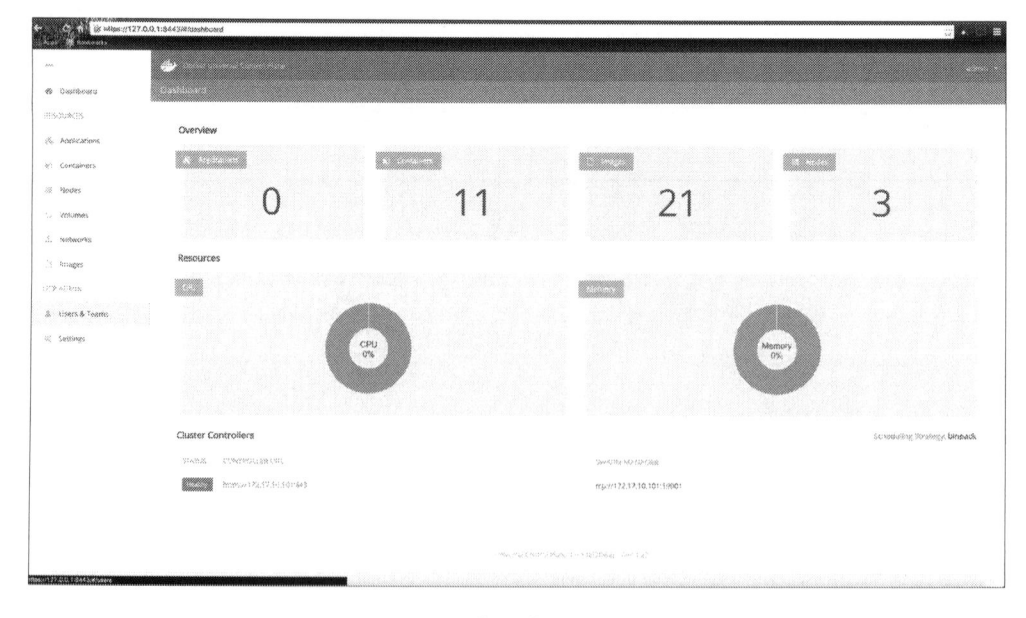

In this book, we are not going to go into how to administer the cluster through the web UI. I would definitely recommend that you explore this product, as it has some really cool features. All the documentation is available at https://docs.docker.com/ucp/overview/.

Kubernetes

There is massive buzz around Kubernetes at the moment. This is Google's offering to the container world. Kubernetes has some very heavy backers such as Google, CoreOS, and Netflix. Out of all the schedulers that we have looked at, Kubernetes is the most complex and is highly driven by APIs. If you are new to Kubernetes, I would suggest that you read further about the product (http://kubernetes.io/). We are going to first look at the architecture of Kubernetes, as there are a few moving parts. Then, we will write our module, and we are going to build Kubernetes completely with containers.

The architecture

We will be building Kubernetes on a single node. The reason for this is that it will cut out some of the complexity on how to use flannel over the Docker bridge. This module will give you a good understanding of how Kubernetes works using more advanced Puppet techniques. If you want to take your knowledge further after you have mastered this chapter, I would recommend that you check out the module on the forge at https://forge.puppetlabs.com/garethr/kubernetes. This module really takes Puppet and Kubernetes to the next level.

So what we are going to code looks like the following diagram:

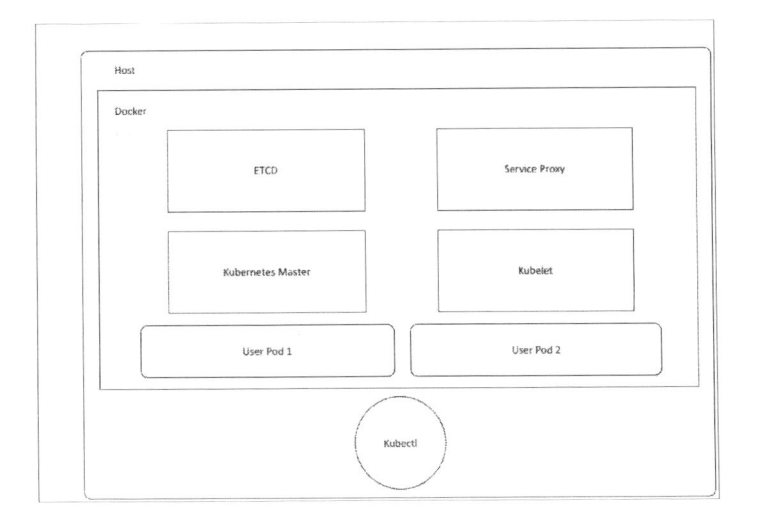

Container Schedulers

As you can see, we have a container that runs **etcd** (to read more about etcd, go to https://coreos.com/etcd/docs/latest/). etcd is similar to Consul, which we are familiar with. In the next few containers, we are going to use hyperkube (https://godoc.org/k8s.io/kubernetes/pkg/hyperkube). This will load balance the required Kubernetes components across multiple containers for us. It seems pretty easy, right? Let's get into the code so that we get a better perspective on all the moving parts.

Coding

We are going to create a new Vagrant repo again. We won't cover how to do that as we have covered it twice in this chapter. If you are unsure, just take a look at the initial part of this chapter.

Once we have created our Vagrant repo, let's open our `servers.yaml` file, as follows:

```yaml
---
-
  box: scottyc/centos-7-puppet-kernel-4-4
  cpu: 1
  ip: "172.17.9.101"
  name: kubernetes
  forward_ports:
    - { guest: 80, host: 8080 }
  ram: 2048
  shell_commands:
    - { shell: 'systemctl stop firewalld && systemctl disable firewalld'}
    - { shell: 'yum install -y wget git lvm2 device-mapper-libs' }
    - { shell: 'echo -e "172.17.9.101 kubernetes">/etc/hosts'}
    - { shell: '/opt/puppet/bin/gem install r10k && ln -s /opt/puppet/bin/r10k /usr/bin/r10k || true'}
    - { shell: 'cp /home/vagrant/kubernetes/Puppetfile /tmp && cd /tmp && r10k puppetfile install --verbose' }
    - { shell: 'cp /home/vagrant/kubernetes/modules/* -R /tmp/modules' }
```

As you can see, there is nothing special there that we have not covered in this book. There's just a single node that we mentioned earlier, `kubernetes`. The next file we will look at is our Puppetfile. We will, of course, need our Docker module, `stdlib`, and lastly, `wget`. We need `wget` to get `kubectl`:

```ruby
#!/usr/bin/ruby env

require "socket"
require 'resolv'

forge 'http://forge.puppetlabs.com'

mod 'puppetlabs/stdlib'
mod 'garethr/docker', :git => "https://github.com/scotty-c/garethr-docker.git"
mod 'maestrodev/wget'
```

That is all the plumbing that we need to set up for our repo. Let's create a new module called `<AUTHOR>-kubernetes_docker`. Once it is created, we will move our module to the `modules` directory in the root of our Vagrant repo.

We are going to create two new folders in our module. The first will be the `templates` folder, and the other folder will be the `lib` directory. We will get to the `lib` directory toward the end of our coding. The first file we will create and edit is `docker-compose.yml.erb`. The reason for this is that it is the foundation of our module. We will add the following code to it:

```
---
etcd:
    image: gcr.io/google_containers/etcd:2.2.1
    net: host
    command: ['/usr/local/bin/etcd', '--bind-addr=0.0.0.0:4001', '--data-dir=/var/etcd/data']
    restart: always
hyperkube:
    image: gcr.io/google_containers/hyperkube:v1.1.8
    volumes:
        - /:/rootfs:ro
        - /sys:/sys:ro
        - /dev:/dev
        - /var/lib/docker/:/var/lib/docker:ro
        - /var/lib/kubelet/:/var/lib/kubelet:rw
        - /var/run:/var/run:rw
        - /kubeconfig:/etc/kubernetes/manifests:ro
    net: host
    pid: host
    privileged: true
    restart: always
    command: ['/hyperkube', 'kubelet', '--containerized', '--address=0.0.0.0', "--api-servers=http://<%= @master_ip %>:8080", '--config=/etc/kubernetes/manifests']
proxy:
    image: gcr.io/google_containers/hyperkube:v1.1.8
    net: host
    pid: host
    privileged: true
    restart: always
    command: ['/hyperkube', 'proxy', "--master=http://<%= @master_ip %>:8080", '--v=2']
```

Let's break this file down into three chunks, as there is a lot going on there. The first block of code is going to set up our etcd cluster. You can see from the screenshot name that we are using Google's official images, and we are using etcd version 2.2.1. We are giving the container access to the host network. Then, in the command resource, we pass some arguments to etcd as it starts.

The next container we create is hyperkube. Again, it is an official Google image. Now, we are giving this container access to a lot of host volumes, the host network, and the host process, making the container privileged. This is because the first container will bootstrap Kubernetes and it will spawn more containers running the various Kubernertes components. Now, in the command resource, we are again passing some arguments for hyperkube. The two major ones we need to worry about are the API server address and config manifests. You will note that we have a mapped folder from `/kubeconfig:/etc/kubernetes/manifests:ro`. We are going to modify our manifest file to make our Kubernetes environment available to the outside world. We will get to that next. But, we will finish looking at the code in this file first.

Container Schedulers

The last container and the third block of code is going to set up our service proxy. We are going to give this container access to the host network and process. In the command resource, we are going to specify that this container is a proxy. The next thing to take notice of is that we specify where the proxy can find the API. Now, let's create the next file, `master.json.erb`. This is the file that hyperkube will use to schedule all the Kubernetes components, which are as follows:

```
{
"apiVersion": "v1",
"kind": "Pod",
"metadata": {"name":"k8s-master"},
"spec":{
  "hostNetwork": true,
  "containers":[
    {
      "name": "controller-manager",
      "image": "gcr.io/google_containers/hyperkube:v1.1.8",
      "command": [
        "/hyperkube",
        "controller-manager",
        "--master=<%= @master_ip %>:8080",
        "--v=2"
      ]
    },
    {
      "name": "apiserver",
      "image": "gcr.io/google_containers/hyperkube:v1.1.8",
      "command": [
        "/hyperkube",
        "apiserver",
        "--portal-net=10.0.0.1/24",
        "--address=<%= @master_ip %>",
        "--etcd_servers=http://<%= @master_ip %>:4001",
        "--cluster_name=kubernetes",
        "--v=2"
      ]
    },
    {
      "name": "scheduler",
      "image": "gcr.io/google_containers/hyperkube:v1.1.8",
      "command": [
        "/hyperkube",
        "scheduler",
        "--master=<%= @master_ip %>:8080",
        "--v=2"
      ]
    }
  ]
}
```

As you can see, we have defined three more containers. This is the first time we will define a Kubernetes pod (`http://kubernetes.io/docs/user-guide/pods/`). A pod is a group of containers that creates an application. It is similar to what we have done with Docker Compose. As you can see, we have changed all the IP addresses to the `<%= @master_ip %>` parameter. We will create four new files: `apps.pp`, `config.pp`, `install.pp`, and `params.pp`.

We will now move on to our files in the `modules` manifest directory. Now, strap yourselves in, as this is where the magic happens. Well, that's not true. The magic happens here and in our `lib` directory. We will need to write some custom types and providers for Puppet to be able to control Kubernertes as `kubectl` is the user interface (for types, visit https://docs.puppetlabs.com/guides/custom_types.html, and for providers, visit https://docs.puppetlabs.com/guides/provider_development.html).

Let's start with our `init.pp` file, which is as follows:

```
class kubernetes_docker (

    $master_ip = $kubernetes_docker::params::master_ip,

) inherits kubernetes_docker::params {

    include kubernetes_docker::install
    contain kubernetes_docker::config
    contain kubernetes_docker::apps
}
```

As you can see, there is not much in this file. We are going to use our `init.pp` file to control the order in which classes are executed. We are also declaring param `<%= @master_ip %>`. We will now move on to our `install.pp` file, as follows:

```
class kubernetes_docker::install {

    package { 'device-mapper-libs':
        ensure => installed,
    }

    class { 'docker':
        tcp_bind    => 'tcp://127.0.0.1:4243',
        socket_bind => 'unix:///var/run/docker.sock',
        require     => Package['device-mapper-libs']
    } ->

    file { '/kubeconfig':
        ensure => directory,
        group  => 'docker',
        mode   => '0770',
    } ->

    file { '/kubeconfig/master.json':
        ensure  => file,
        content => template('kubernetes_docker/master.json.erb'),
        mode    => '0755',
    } ->

    file { '/root/docker-compose.yml':
        ensure  => file,
        content => template('kubernetes_docker/docker-compose.yml.erb'),
    } ->

    docker_compose { kubernetes :
        ensure => present,
        source => '/root',
        scale  => ['1', '1', '1']
    }
}
```

Container Schedulers

In this file, we install Docker as we did before. We will place our two templates that we created earlier. Then, we will run Docker Compose to bring up our cluster. Now, we will move on to `config.pp`, as follows:

```
class kubernetes_docker::config {

  wget::fetch { 'kubectl':
    source      => 'https://storage.googleapis.com/kubernetes-release/release/v1.8/bin/linux/amd64/kubectl',
    destination => '/usr/bin/kubectl',
    require     => Class['kubernetes_docker::install']
  } ->

  file { '/usr/bin/kubectl':
    mode => '0777',
  } ->

  kubectl_config {'default-cluster':
    cluster      => 'kubernetes',
    kube_context => 'puppet',
  }
}
```

The first thing that we declare is that we want `wget` to be our `kubectl` client, and we place it at `/usr/bin/` (http://kubernetes.io/docs/user-guide/kubectl/kubectl/). You really need to understand what this interface does; otherwise, you might get a bit lost from here. So, I suggest that you have a fairly good idea of what kubectl is and what it is capable of. Next, we will make it executable and available for all our users. Now, this last piece of code does not make sense, as we have not called the `kubectl_config` class yet:

```
kubectl_config {'default-cluster':
  cluster      => 'kubernetes',
  kube_context => 'puppet',
}
```

We now need to jump to our `lib` directory. This first thing we will do is create all our folders that we need. The first folder that we will create is `puppet` under the `lib` directory. We will look at our custom types first. We will create a folder called `type` under `puppet`. The following screenshot will give you a visualization of the structure:

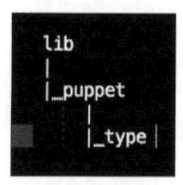

```
lib
|
|_puppet
    |
    |_type
```

Under the `type` folder, we will create a file called `kubectl_config.rb`. In that file, we will add the new parameters of `type` as follows:

```
Puppet::Type.newtype(:kubectl_config) do
  @doc = "configures kubernetes cluster"

  ensurable do
    defaultvalues
    defaultto :present
  end

  newparam(:name, :namevar => true) do
    desc "resource name"
  end

  newparam(:cluster) do
    desc "cluster nickname"
  end

  newparam(:kube_context) do
    desc "the context to add to the cluster"
  end
end
```

Let me explain what is happening here. In the first line, we are going to declare our new type, `kubectl_config`. We are then going to set the default value of the new type when it is declared as `present`. We are going to declare three values to our type, `name`, `cluster`, and `kube_context`. These are all settings that we will add to our `config` file that `kubectl` will use when we interface with it. Now, we will create a folder under the `lib` directory called `provider`. Then, under that, we will create a folder with the same name as our custom type, `kubectl_config`. Inside that folder, we will create a file called `ruby.rb`. In this file, we will put the Ruby code that provides logic to our type as follows:

```ruby
require 'socket'
require 'resolv'
require 'fileutils'

Puppet::Type.type(:kubectl_config).provide(:ruby) do
  desc "support for configuring a kubernetes cluster"

  mk_resource_methods

  commands :kubectl => "kubectl"

  def interface
    hostname = Socket.gethostname
    IPSocket.getaddress(hostname)
  end

  def kube_conf_server
    run = ['config','set-cluster', "#{resource[:cluster]}", "--server=http://#{interface}:8080", '--insecure-skip-tls-verify=true']
  end

  def kube_conf_context
    run = ['config','set-context', "#{resource[:kube_context]}", "--cluster=#{resource[:cluster]}"]
  end

  def kube_conf_use
    run = ['config','use-context', "#{resource[:kube_context]}"]
  end

  def exists?
    Puppet.info("checking if kubectl is configured")
    File.exist?('/.kube/config')
  end

  def create
    Puppet.info("configuring kubernetes cluster")
    kubectl *kube_conf_server
    kubectl *kube_conf_context
    kubectl *kube_conf_use
    FileUtils.ln_s('/.kube', '/root/.kube')
  end

  def destroy
    Puppet.info("removing kubectl config")
    FileUtils.rm_rf('/.kube')
  end
end
```

Container Schedulers

A provider needs to have three methods for Puppet to be able to run the code. They are `exsists?`, `create`, and `destroy`. These are pretty easy to understand. The `exists?` method checks whether the type has already been executed by Puppet, `create` runs the type, and `destroy` is invoked when the type is set to `absent`.

We are now going to work through the file, from top to bottom. We need to first load a few Ruby libraries for some of our methods. We will then tie this provider to our type. The next thing that we need to declare is the `kubectl` executable.

Now we will write our first method, `interface`. This will get the IP address from the hostname of the box. We will then create three more methods. We will also create an array and add all our configuration to them. You will note that we are mapping our parameters from our type in the arrays.

In our `exists?` method, we will check for our `kubectl` config file.

In our `create` method, we are calling our `kubectl` executable and then passing our arrays as arguments. We will then link our `config` file to roots' home directory (this is fine for our lab. In a production environment, I would use a named user account).

Lastly, we will remove the `config` file if the type is set to `absent`. We will now go back to our manifests directory and look at our last file, which is `apps.pp`:

```
class kubernetes_docker::apps {

  kubernetes_run { 'nginx':
    image   => 'nginx',
    port    => '80',
    require => Class['kubernetes_docker::config']
  }
}
```

In this file, we are going to run a container application on our Kubernetes cluster. Again, we will write another custom type and provider. Before we get to that, we should look at the code in this class. As you can see, our type is called `kubernetes_run`. We can see that our service is named `nginx`, the Docker image we will pull is `nginx`, and we will then expose port `80`.

Let's go back to our `lib` directory. We will then create a file in our `type` folder called `kubernetes_run.rb`. In this file, we will set up our custom type as we did earlier:

Chapter 7

```ruby
Puppet::Type.newtype(:kubernetes_run) do
    @doc = "configures kubernetes applications to run on our cluster"

    ensurable do
      defaultvalues
      defaultto :present
    end

    newparam(:service_name, :namevar => true) do
      desc "resource name"
    end

    newparam(:image) do
      desc "the docker image to use"
    end

    newparam(:port) do
      desc "the port to expose"
    end
end
```

As you can see, we are mapping the same parameters that we had in our `apps.pp` file. We will then create a folder under the `provider` folder with the same name as our `kubernetes_run` type. Again, in the newly created directory, we will create a file called `ruby.rb`. It will have the code shown in the following screenshot:

```ruby
Puppet::Type.type(:kubernetes_run).provide(:ruby) do
  desc "support for configuring a kubernetes cluster"

  mk_resource_methods

  commands :kubectl => "kubectl"
  commands :docker => "docker"

  def kube_run
    run = ['run', "#{(resource[:service_name])}", "--image=#{(resource[:image])}", "--port=#{(resource[:port])}"]
  end

  def kube_expose
    run = ['expose', 'rc', "#{(resource[:service_name])}", "--port=#{(resource[:port])}", "--external-ip=#{interface}"]
  end

  def exists?
    Puppet.info("checking kubectl if svc #{(resource[:service_name])} is configured")
    begin
      exists_args = ['get', 'svc']
      run = kubectl *exists_args
      run.match("#{(resource[:service_name])}")
    rescue => e
      return false
    end
  end

  def create
    Puppet.info("running #{(resource[:service_name])} on kubernetes cluster")
    begin
      args = ['get', 'nodes']
      kubectl *args
    rescue => e
      retry
    ensure
      kubectl *kube_run
      kubectl *kube_expose
    end
  end

  def destroy
    Puppet.info("removing application")
    destroy_args = ['rm', '-f', "#{(resource[:service_name])}"]
    docker *destroy_args
  end
end
```

Container Schedulers

In this file, we are going to add two commands this time: the first is `kubectl` and the second is `docker`. We will create two methods, again with arrays that map the values from our type.

Now, let's look at our `exists?` method. We are going to pass an array as an argument to `kubectl` to check whether the service exists. We are then going to catch the error if `kubectl` throws an error with the request and returns `false`. This is used if there are no services deployed on the cluster.

In our `create` method, we will first pass an array to `kubectl` to get the nodes in the cluster. We are using this as an arbitrary command to make sure that the cluster is up. Under that, we will capture the error and retry the command until it is successful. Once it is successful, we will deploy our container with the `ensure` resource.

In the `destroy` method, we will use `docker` to remove our container.

Now we have all our coding done. We just need to add our class to our node by editing our `default.pp` file in the `manifests` folder in the root of our Vagrant repo as follows:

```
1  node 'kubernetes' {
2
3      include kubernetes_docker
4
5  }
6
```

Now, let's open our terminal and change the directory to the root of our Vagrant repo and issue the `vagrant up` command. Once Puppet has completed its run, our terminal should look like the following screenshot:

```
kubernetes: Running: inline script
kubernetes: Running provisioner: puppet...
kubernetes: Running Puppet with environment production...
kubernetes: Info: Loading facts
kubernetes: Notice: Compiled catalog for localhost in environment production in 1.05 seconds
kubernetes: Info: Applying configuration version '1458034261'
kubernetes: Notice: /Stage[main]/Docker::Repos/Yumrepo[docker]/ensure: created
kubernetes: Info: changing mode of /etc/yum.repos.d/docker.repo from 600 to 644
kubernetes: Notice: /Stage[main]/Docker::Install/Package[docker]/ensure: created
kubernetes: Notice: /Stage[main]/Docker::Service/File[/etc/sysconfig/docker-storage-setup]/ensure: created
kubernetes: Info: /Stage[main]/Docker::Service/File[/etc/sysconfig/docker-storage-setup]: Scheduling refresh of Service[docker]
kubernetes: Notice: /Stage[main]/Docker::Service/File[/etc/systemd/system/docker.service.d]/ensure: created
kubernetes: Notice: /Stage[main]/Docker::Service/File[/etc/systemd/system/docker.service.d/service-overrides.conf]/ensure: created
kubernetes: Info: /Stage[main]/Docker::Service/File[/etc/systemd/system/docker.service.d/service-overrides.conf]: Scheduling refresh of Exec[docker-systemd-reload-before-service]
kubernetes: Notice: /Stage[main]/Docker::Service/Exec[docker-systemd-reload-before-service]/returns: executed successfully
kubernetes: Notice: /Stage[main]/Docker::Service/Exec[docker-systemd-reload-before-service]: Triggered 'refresh' from 1 events
kubernetes: Notice: /Stage[main]/Docker::Service/File[/etc/sysconfig/docker-storage]/ensure: created
kubernetes: Info: /Stage[main]/Docker::Service/File[/etc/sysconfig/docker-storage]: Scheduling refresh of Service[docker]
kubernetes: Notice: /Stage[main]/Docker::Service/File[/etc/sysconfig/docker]/ensure: created
kubernetes: Info: /Stage[main]/Docker::Service/File[/etc/sysconfig/docker]: Scheduling refresh of Service[docker]
kubernetes: Notice: /Stage[main]/Docker::Service/Service[docker]/ensure: ensure changed 'stopped' to 'running'
kubernetes: Info: /Stage[main]/Docker::Service/Service[docker]: Unscheduling refresh on Service[docker]
kubernetes: Notice: /Stage[main]/Docker::Compose/Exec[Install Docker Compose 1.6.0]/returns: executed successfully
kubernetes: Notice: /Stage[main]/Docker::Compose/File[/usr/local/bin/docker-compose-1.6.0]/mode: mode changed '0644' to '0755'
kubernetes: Notice: /Stage[main]/Docker::Compose/File[/usr/local/bin/docker-compose]/ensure: created
kubernetes: Notice: /Stage[main]/Kubernetes_docker::Install/File[/kubeconfig]/ensure: created
kubernetes: Notice: /Stage[main]/Kubernetes_docker::Install/File[/kubeconfig/master.json]/ensure: defined content as '{md5}f3fd255926e4113cnf30af4e65906a5e'
kubernetes: Notice: /Stage[main]/Kubernetes_docker::Install/File[/root/docker-compose.yml]/ensure: defined content as '{md5}08cdcd9d97c6db3cbeca1d3d87175du6'
kubernetes: Info: Checking if docker-compose.yml exists
kubernetes: Info: bring up containers
kubernetes: Notice: /Stage[main]/Kubernetes_docker::Install/Docker_compose[kubernetes]/ensure: created
kubernetes: Notice: /Stage[main]/Kubernetes_docker::Config/Wget::Fetch[kubectl]/Exec[wget-kubectl]/returns: executed successfully
kubernetes: Notice: /Stage[main]/Kubernetes_docker::Config/File[/usr/bin/kubectl]/mode: mode changed '0644' to '0777'
kubernetes: Info: checking if kubectl is configured
kubernetes: Info: configuring kubernetes cluster
kubernetes: Notice: /Stage[main]/Kubernetes_docker::Config/Kubectl_config[default-cluster]/ensure: created
kubernetes: Info: checking kubectl if svc nginx is configured
kubernetes: Info: running nginx on kubernetes cluster
kubernetes: Notice: /Stage[main]/Kubernetes_docker::Apps/Kubernetes_run[nginx]/ensure: created
kubernetes: Notice: Finished catalog run in 165.10 seconds
```

We will now log in to our vagrant box by issuing the `vagrant ssh` command and then `sudo -i` and change to root. Once we are root, we will look at our service on our cluster. We will do this by issuing the `kubectl get svc` command as follows:

```
[root@kubernetes ~]# kubectl get svc
NAME         CLUSTER_IP   EXTERNAL_IP    PORT(S)   SELECTOR    AGE
kubernetes   10.0.0.1     <none>         443/TCP   <none>      1h
nginx        10.0.0.50    172.17.9.101   80/TCP    run=nginx   1h
[root@kubernetes ~]#
```

As you can see, we have two services running our cluster: `Kubernetes` and `nginx`. If we go to our web browser and go to the address we gave to the second network interface, `http://172.17.9.101`, we will get the following nginx default page:

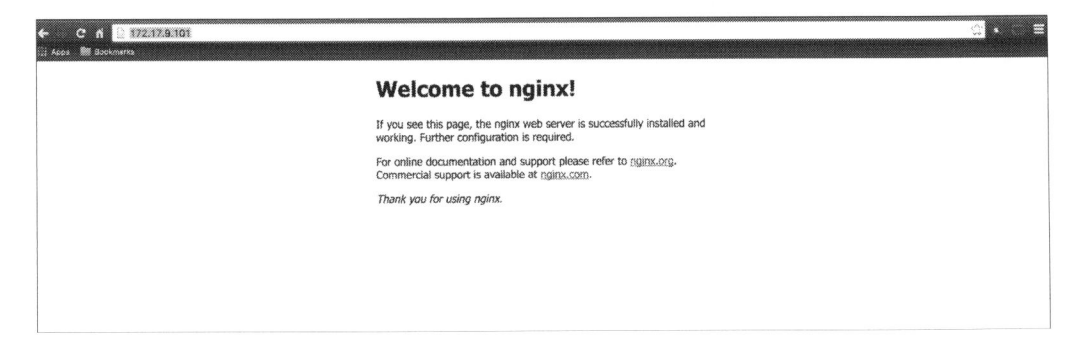

Welcome to nginx!

If you see this page, the nginx web server is successfully installed and working. Further configuration is required.

For online documentation and support please refer to nginx.org.
Commercial support is available at nginx.com.

Thank you for using nginx.

Now, our cluster is running successfully with our `nginx` service.

[153]

Summary

We covered my favorite three container schedulers; each one of them has their own pros and cons. Now, you have the knowledge and the required code to give all three a really good test run. I would suggest that you do so so that you can make the right choice when choosing the design in your environment.

8
Logging, Monitoring, and Recovery Techniques

In this chapter, we are going to look into one of our schedulers and look at wrapping some more operational tasks around it. So far, in this book, we have covered more glamorous subjects; however, monitoring, logging, and automated recovery are just as important. We want to take this knowledge and make it work in the real world. From there, we can start to see the benefits to both the development and operations teams. We are going to use Docker Swarm as our scheduler in this chapter. For logging, we will use the ELK stack, and for monitoring, we will use Consul. Since Docker Swarm version 1.1.3, there are some cool features that will help us use recovery, so we will look at them. We will cover the following topics in this chapter:

- Logging
- Monitoring
- Recovery techniques

Logging
The importance of logging is second to none in a solution. If we need to debug any issues with any code/infrastructure, the logs are the first place to go. In the container world, this is no different. In one of the previous chapters, we built the ELK stack. We are going to use that again to process all the logs from our containers. In this solution, we will use a fair chunk of the knowledge that we got so far. We will use a scheduler, a Docker network, and lastly, service discovery with Consul. So, let's look at the solution, and like we did in the past chapters, we will get to coding.

The solution

As I mentioned in the introduction of this chapter, we will be using Docker Swarm for this solution. The reason for this choice is that I want to highlight some of the features of Swarm as it has come on leaps and bounds in the last few releases. For the logging portion of this chapter, we are going to deploy our three containers and let Swarm schedule them. We will use a combination of the Docker networking DNS and our service discovery with Consul to tie everything together. In Swarm, we will use the same servers as we did in the last chapter: three member nodes and two replicated masters. Each node will be a member of our Consul cluster. We will again use Puppet to install Consul on the host system natively.

The code

In this chapter, we will build on the code that we used in the last chapter for Docker Swarm. So, we will go through the plumbing of the Vagrant repo fairly swiftly, only calling out the differences from the last chapter. We are going to create a new Vagrant repo again for this chapter. You should be a master at this by now. Once the new repo is set up, open the `servers.yaml` file. We will add the following code to it:

Code for the severs.yaml file

As you can see, it's not that different from the last chapter. There is one call out. We have added a new line to each server - `{ shell: 'echo -e "PEERDNS=no\nDNS1=127.0.0.1\nDNS2=8.8.8.8">>/etc/sysconfig/network-scripts/ifcfg-enp0s8 && systemctl restart network' }`. We will add this as the server is multihomed. We will want to make sure that we are resolving DNS correctly on each interface.

Logging, Monitoring, and Recovery Techniques

The next thing we will look at is the puppetfile, which is as follows:

```ruby
#!/usr/bin/ruby env

require "socket"
$hostname = Socket.gethostname

forge 'http://forge.puppetlabs.com'


mod 'puppetlabs/stdlib'
mod 'puppetlabs/vcsrepo'
mod 'nanliu/staging'
mod 'KyleAnderson/consul'
mod 'scottyc/docker_swarm'
mod 'scottyc/golang'
mod 'garethr/docker', :git => "https://github.com/scotty-c/garethr-docker.git"
mod 'stankevich/python'
mod 'stahnma/epel'
mod 'maestrodev/wget'
```

As you can see, there are no new changes compared with the last chapter. So, let's move to our Hiera file located at `heiradata/global.yml` in the root of our module:

```yaml
docker_swarm::swarm_version: v1.1.3
docker_swarm::backend: consul
docker_swarm::backend_ip: 172.17.8.101
docker_swarm::backend_port: 8500
docker_swarm::advertise_int: enp0s8
consul::version: 0.6.3
```

Again, we are setting the Swarm version to `v1.1.3`, and the backend is set to `consul`. We set the IP address of the first node in the Consul cluster, and we set the Consul port to `8500`. We will set the `swarm` interface that we will advertise from, and last but not least, we will set our Consul version to `0.6.3`.

Now, we will create our module. We will again call the `config` module. Once you have created your `<AUTHOR>-config` module, move it to the `modules` folder in the root of your Vagrant repo.

Now that we have our module, let's add our code to it. We will need to create the following files in the `manifests` directory: `compose.pp`, `consul_config`, `dns.pp`, `run_containers.pp`, and `swarm.pp`. We have no `params.pp` as we are using Hiera in this example.

So, let's go through the files in an alphabetical order. In our `compose.pp` file, we will add the following code:

```
class config::compose {

  if $hostname =~ /^swarm-master*/ {

    notice ["This server is the Swarm Manager."]

  }

  else {

    file { '/root/docker-compose.yml':
      ensure  => file,
      content => template("config/registrator.yml.erb"),
    } ->

    docker_compose {'swarm app':
      ensure => present,
      source => '/root',
      scale  => ['1']
    }
  }
}
```

Logging, Monitoring, and Recovery Techniques

As you can see from the code, we are adding our `docker-compose.yml` file to any node that is not a swarm master. We will come back to the `docker compose.yml` file when we look at the `templates` directory. The next file is `consul_config.pp`, which is as follows:

```
class config::consul_config {

  if $hostname =~ /^*101*$/ {

    class { 'consul':
      config_hash => {
        'datacenter'        => 'dev',
        'data_dir'          => '/opt/consul',
        'ui_dir'            => '/opt/consul/ui',
        'bind_addr'         => $::ipaddress_enp0s8,
        'client_addr'       => '0.0.0.0',
        'node_name'         => "$::hostname",
        'advertise_addr'    => '172.17.8.101',
        'bootstrap_expect'  => '1',
        'server'            => true
      }
    }
  }
  else {

    class { 'consul':
      config_hash => {
        'bootstrap'      => false,
        'datacenter'     => 'dev',
        'data_dir'       => '/opt/consul',
        'ui_dir'         => '/opt/consul/ui',
        'bind_addr'      => $::ipaddress_enp0s8,
        'client_addr'    => '0.0.0.0',
        'node_name'      => "$::hostname",
        'advertise_addr' => $::ipaddress_enp0s8,
        'start_join'     => ['172.17.8.101','172.17.8.103','172.17.8.103'],
        'server'         => false
      }
    }
  }

  consul::service { 'docker-service':
    checks => [
      {
        script   => 'service docker status',
        interval => '10s',
        tags     => ['docker-service']
      }
    ],
    address => $::ipaddress_enp0s8,
  }

}
```

In this file, we are declaring our Consul cluster and defining the bootstrap server. This should look familiar as it is the same code that we used in the last chapter. The next file is `dns.pp`, which is given as follows:

```puppet
class config::dns {

  package { 'bind':
    ensure => present
  } ->

  file { '/etc/named.conf':
    ensure  => present,
    content => template("config/named.conf.erb"),
    mode    => '0644',
    owner   => 'root',
    group   => 'root',
    require => Package['bind'],
  } ~>

  file { '/etc/named/consul.conf':
    ensure  => present,
    content => template("config/consul.conf.erb"),
    mode    => '0644',
    owner   => 'root',
    group   => 'root',
    require => Package['bind'],
  } ~>

  service { 'named':
    ensure  => running,
    enable  => true,
    require => File['/etc/named.conf'],
  }
}
```

Again, this code should look familiar to you, as we have used it before in the last chapter. Just to recap, this is setting and configuring our bind package to use Consul as the DNS server. The next file we will look at is `init.pp`:

```puppet
# == Class: config
#
#
class config(

  $consul_ip = "$::ipaddress_enp0s8",

){

  include config::consul_config
  contain config::swarm
  contain config::compose
  contain config::dns
  contain config::run_containers

  Class['config::swarm'] -> Class['config::compose'] -> Class['config::dns'] -> Class['config::run_containers']
}
```

Logging, Monitoring, and Recovery Techniques

In the `init.pp` file, we are just ordering our classes within our module. We will now move on to `run_containers.pp`. This is where we will schedule our ELK containers across the swarm cluster:

```puppet
class config::run_containers {

  if $hostname =~ /swarm-master-02/ {

    swarm_run {'logstash':
      ensure  => present,
      image   => 'scottyc/logstash',
      network => 'swarm-private',
      ports   => ['9998:9998', '9999:9999/udp', '5000:5000', '5000:5000/udp'],
      env     => ['ES_HOST=elasticsearch', 'ES_PORT=9200'],
      command => 'logstash -f /opt/logstash/conf.d/logstash.conf --debug',
      require => Class['config::swarm']
    }

    swarm_run {'elasticsearch':
      ensure     => present,
      image      => 'elasticsearch:2.1.0',
      network    => 'swarm-private',
      volumes    => ['/etc/esdata:/usr/share/elasticsearch/data'],
      command    => 'elasticsearch -Des.network.host=0.0.0.0',
      log_driver => 'syslog',
      log_opt    => 'syslog-address=tcp://logstash-5000.service.consul:5000',
      depends    => 'logstash',
      require    => Class['config::swarm']
    }

    swarm_run {'kibana':
      ensure     => present,
      image      => 'kibana:4.3.0',
      network    => 'swarm-private',
      ports      => ['80:5601'],
      env        => ['ELASTICSEARCH_URL=http://elasticsearch:9200'],
      log_driver => 'syslog',
      log_opt    => 'syslog-address=tcp://logstash-5000.service.consul:5000',
      depends    => 'logstash',
      require    => Class['config::swarm']
    }
  }
}
```

Let's have a look at this in detail, as there is a lot of new code here. The first declaration that we will use is to schedule the containers from the second Swarm master.

The next block of code will configure our `logstash` container. We will need to first have these containers here in this example, as we are using them as the syslog server. If at the time of spawning the containers they can't connect to logstash on TCP port `5000`, the build of the container will fail. So, let's move on to the configuration of `logstash`. We will use the container that I put in, as it is the official container with the `logstash.conf` file that we already added. We will then add `logstash` to our internal `swarm-private` Docker network and expose all the ports for `logstash` on all networks. So, we can pipe logs from anywhere to it. After this, we will set the location of `elasticsearch` and then we will give the command to start.

Logstash

In the second block of code, we will install and configure `elasticsearch`. We will use the official `elasticsearch` container (version 2.1.0). We will only add `elasticsearch` to our private Docker network, `swarm-private`. We will make the data persistent by declaring a volume mapping. We will set the command with arguments to start `elasticsearch` with the command value. Next, we will set the log drive to syslog and the syslog server to `tcp://logstash-5000.service.consul:5000`. Note that we are using our Consul service discovery address as we are exposing `logstash` on the external network. Lastly, we set the dependency on `logstash`. As I mentioned earlier, the syslog server needs to be available at the time this container spawns, so we need `logstash` to be there prior to either this container or `kibana`. Talking of Kibana, let's move on to our last block of code.

In our `kibana` container, we will add the following configuration. First, we will use the official `kibana` image (Version 4.3.0). We will add `kibana` to our `swarm-private` network so it can access our `elasticsearch` container. We will map and expose ports `5601` to `80` on the host network. In the last few lines, we will set the syslog configuration in the same way as we did with `elasticsearch`.

Now, it's time for our last file, `swarm.pp`, which is as follows:

```
class config::swarm {

  class { 'docker_swarm':
    require => Class['config::consul_config']
  }

  docker_network { 'swarm-private':
    ensure  => present,
    create  => true,
    driver  => 'overlay',
    require => Class['config::consul_config']
  }

  if $hostname =~ /^swarm-master*/ {

    swarm_cluster {'cluster 1':
      ensure       => present,
      backend      => 'consul',
      cluster_type => 'manage',
      port         => '8500',
      address      => '172.17.8.101',
      advertise    => $::ipaddress_enp0s8,
      path         => 'swarm',
    }
  }
  else {

    swarm_cluster {'cluster 1':
      ensure       => present,
      backend      => 'consul',
      cluster_type => 'join',
      port         => '8500',
      address      => '172.17.8.101',
      path         => 'swarm'
    }
  }
}
```

Logging, Monitoring, and Recovery Techniques

In this code, we are configuring our Swarm cluster and Docker network.

We will now move to our `templates` folder in the root of our module. We need to create three files. The two files `Consul.conf.erb` and `named.conf.erb` are for our bind config. The last file is our `registrator.yml.erb` Docker compose file. We will add the code to the following files.

Let's first see the code for `consul.conf.erb`, which is as follows:

```
1  zone "consul" IN {
2      type forward;
3      forward only;
4      forwarders { 127.0.0.1 port 8600; };
5  };
```

Now, let's see the code for `named.conf.erb`, which is as follows:

```
1   options {
2       listen-on port 53 { 127.0.0.1; };
3       listen-on-v6 port 53 { ::1; };
4       directory       "/var/named";
5       dump-file       "/var/named/data/cache_dump.db";
6       statistics-file "/var/named/data/named_stats.txt";
7       memstatistics-file "/var/named/data/named_mem_stats.txt";
8       allow-query     { localhost; };
9       recursion yes;
10
11      dnssec-enable no;
12      dnssec-validation no;
13
14      /* Path to ISC DLV key */
15      bindkeys-file "/etc/named.iscdlv.key";
16
17      managed-keys-directory "/var/named/dynamic";
18  };
19
20  include "/etc/named/consul.conf";
```

Finally, let's see the code for `registrator.yml.erb`, which is as follows:

```
1  registrator:
2    image: gliderlabs/registrator
3    net: "host"
4    volumes:
5     - "/var/run/docker.sock:/tmp/docker.sock"
6    command: consul://<%= @consul_ip %>:8500
7
```

All the code in these files should look fairly familiar, as we have used it in previous chapters.

Now, we have just one more configuration before we can run our cluster. So, let's go to our `default.pp` manifest file in the `manifests` folder located in the root of our Vagrant repo.

Now, we will add the relevant node definitions to our manifest file:

```
node 'swarm-101' { include config }
node 'swarm-102' { include config }
node 'swarm-103' { include config }
node 'swarm-master-01' { include config }
node 'swarm-master-02' { include config }
```

We are ready to go to our terminal, where we will change the directory to the root of our Vagrant repo. As so many times that we did before, we will issue the `vagrant up` command. If you have the boxes still configured from the last chapter, issue the `vagrant destroy -f && vagrant up` command.

Once Vagrant is run and Puppet has built our five nodes, we can now open a browser and enter the `http://127.0.0.1:9501/`. We should get the following page after this:

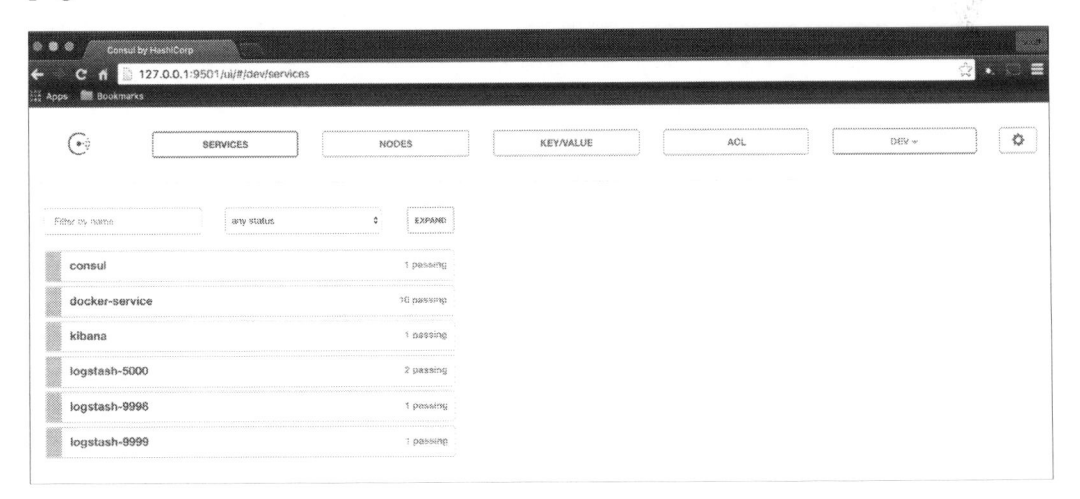

Logging, Monitoring, and Recovery Techniques

As you can see, all our services are shown with green color that displays the state of health. We will now need to find what node our `kibana` container is running on. We will do that by clicking on the **kibana** service.

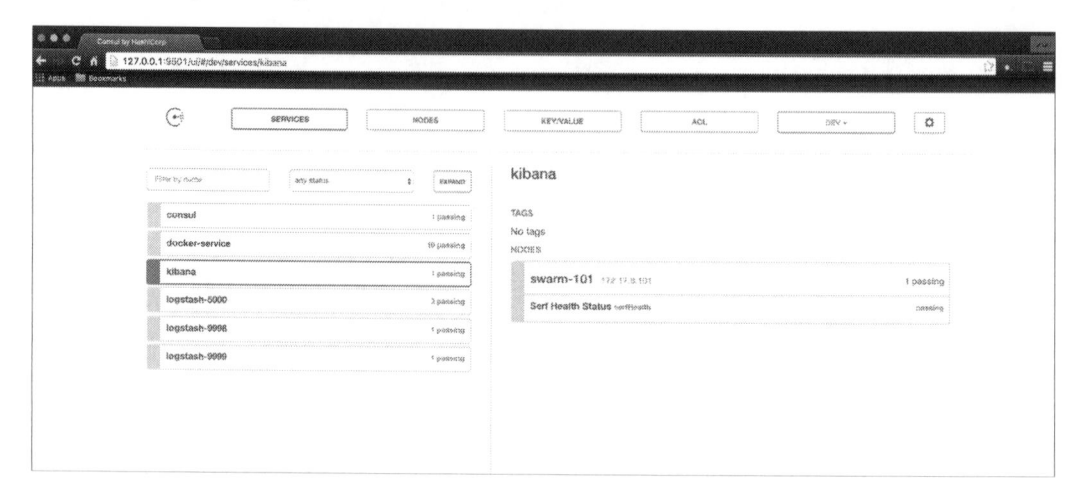

In my example, **kibana** has come up on **swarm-101**. If this is not the same for you, don't worry as the Swarm cluster could have scheduled the container on any of the three nodes. Now, I will open a browser tab and enter the `127.0.0.1:8001/`, as shown in the following screenshot:

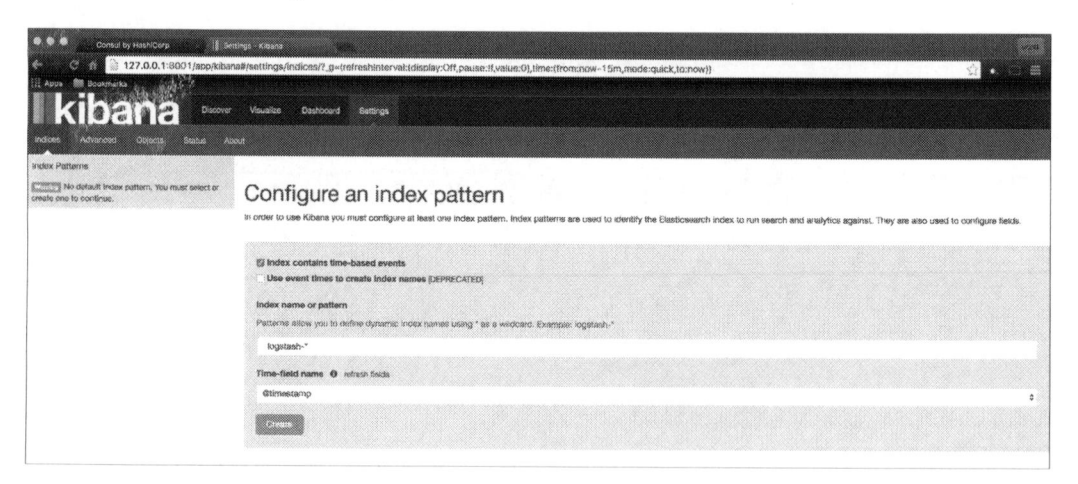

If your host is different, consult the `servers.yaml` file to get the right port.

[166]

We will then create our index and click on the **Discovery** tab, and as you can see in this screenshot, we have our logs coming in:

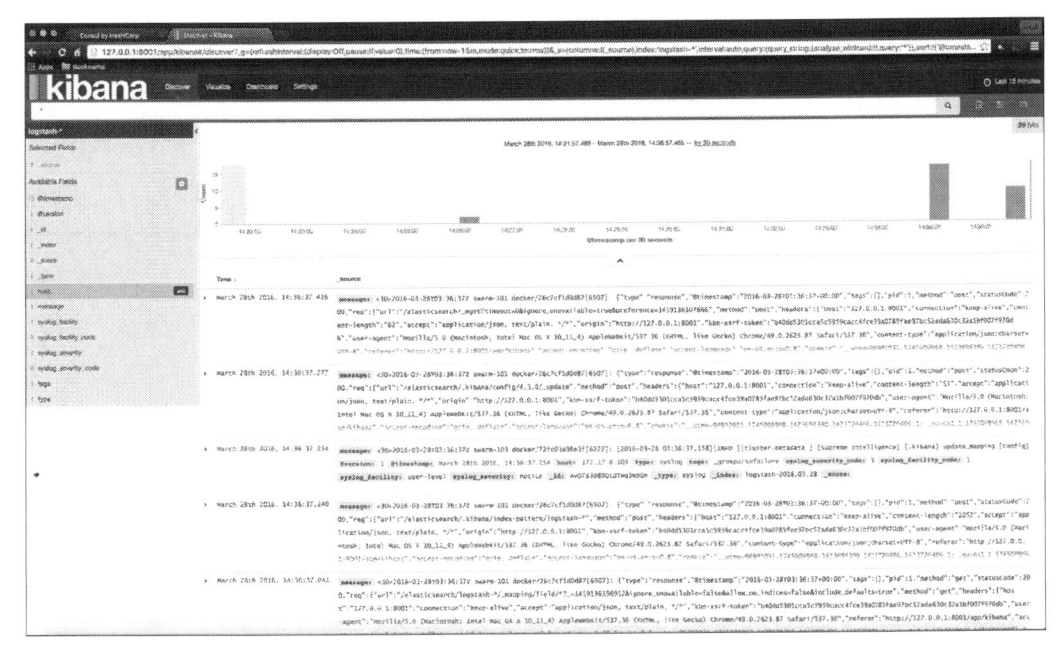

Logs after creating index

Monitoring

In the world of containers, there are a few levels of monitoring that you can deploy. For example, you have your traditional ops monitoring. So your Nagios and Zabbix of the world or even perhaps a cloud solution like Datadog. All these solutions have good hooks into Docker and can be deployed with Puppet. We are going to assume in this book that the ops team has this covered and your traditional monitoring is in place. We are going to look at the next level of monitoring. We will concentrate on the container connectivity and Swarm cluster health. We will do this all in Consul and deploy our code with Puppet.

The reason we are looking at this level of monitoring is because we can make decisions about what Consul is reporting. Do we need to scale containers? Is a Swarm node sick? Should we take it out of the cluster? Now to cover these topics, we will need to write a separate book. I won't be covering these solutions. What we will look at is the first step to get there. Now that the seed has been planted, you will want to explore your options further. The challenge is to change the way we think about monitoring and how it needs reactive interaction from a human, so we can trust our code to make choices for us and to make our solutions fully automated.

Logging, Monitoring, and Recovery Techniques

Monitoring with Consul

One really good thing about using Consul is that Hashicorp did an awesome job at documenting their applications. Consul is no exception. If you would like to read more about the options you have with Consul monitoring, refer to the documentation at `https://www.consul.io/docs/agent/services.html` and `https://www.consul.io/docs/agent/checks.html`. We are going to set up both checks and a service. In the last chapter, we wrote a service with Consul to monitor our Docker service on every node:

```
consul::service { 'docker-service':
  checks  => [
    {
      script   => 'service docker status',
      interval => '10s',
      tags     => ['docker-service']
    }
  ],
  address => $::ipaddress_enp0s8,
}
```

On the Consul web UI, we get the following reading of the Docker service on the node:

```
Service 'docker-service' check   service:docker-service                    passing

NOTES
OUTPUT
  Redirecting to /bin/systemctl status docker.service
  ● docker.service - Docker Application Container Engine
     Loaded: loaded (/usr/lib/systemd/system/docker.service; enabled; vendor preset: disabled)
    Drop-In: /etc/systemd/system/docker.service.d
             └─service-overrides.conf
     Active: active (running) since Mon 2016-03-28 07:33:57 UTC; 8min ago
       Docs: https://docs.docker.com
   Main PID: 6221 (docker)
     CGroup: /system.slice/docker.service
             └─6221 /usr/bin/docker daemon -H tcp://0.0.0.0:2375 -H unix:///var/run/docker.sock --cluste

LOCK SESSIONS
No sessions
```

We are going to roll out all our new checks on both the Swarm masters. The reason for this is that both the nodes are external from the nodes cluster. The abstraction gives us the benefit of not having to worry about the nodes that the containers are running on. You also have the monitoring polling from multiple locations. For example, in AWS, your Swarm masters could be split across multiple AZs. So, if you lose an AZ, your monitoring will still be available.

As we are going to use the logging solution from the example that we covered in the previous section, we will check and make sure that both Logstash and Kibana are available; Logstash on port 5000 and Kibana on port 80.

We are going to add two new service checks to our config module in the `consul_config.pp` file, as follows:

```
if $hostname =~ /^swarm-master*/ {
  consul::check { 'kibana':
    ensure   => present,
    tcp      => 'kibana.service.consul:80',
    interval => '10s',
  }

  consul::check { 'logstash-5000':
    ensure   => present,
    tcp      => 'logstash-5000.service.consul:5000',
    interval => '10s',
  }
```

As you can see, we have set a TCP check for both `kibana` and `logstash`, and we will use the service discovery address to test the connections. We will open our terminal and change the directory to the root of our Vagrant repo.

Logging, Monitoring, and Recovery Techniques

Now, we are going to assume that your five boxes are running. We will issue the command to Vagrant to provision only the two master nodes. This command is `vagrant provision swarm-master-01 && vagrant provision swarm-master-02`. We will then open our web browser and enter `127.0.0.1:9501`. We can then click on **swarm-master-01** or **swarm-master-02**, the choice is up to you. After this, we will get the following result:

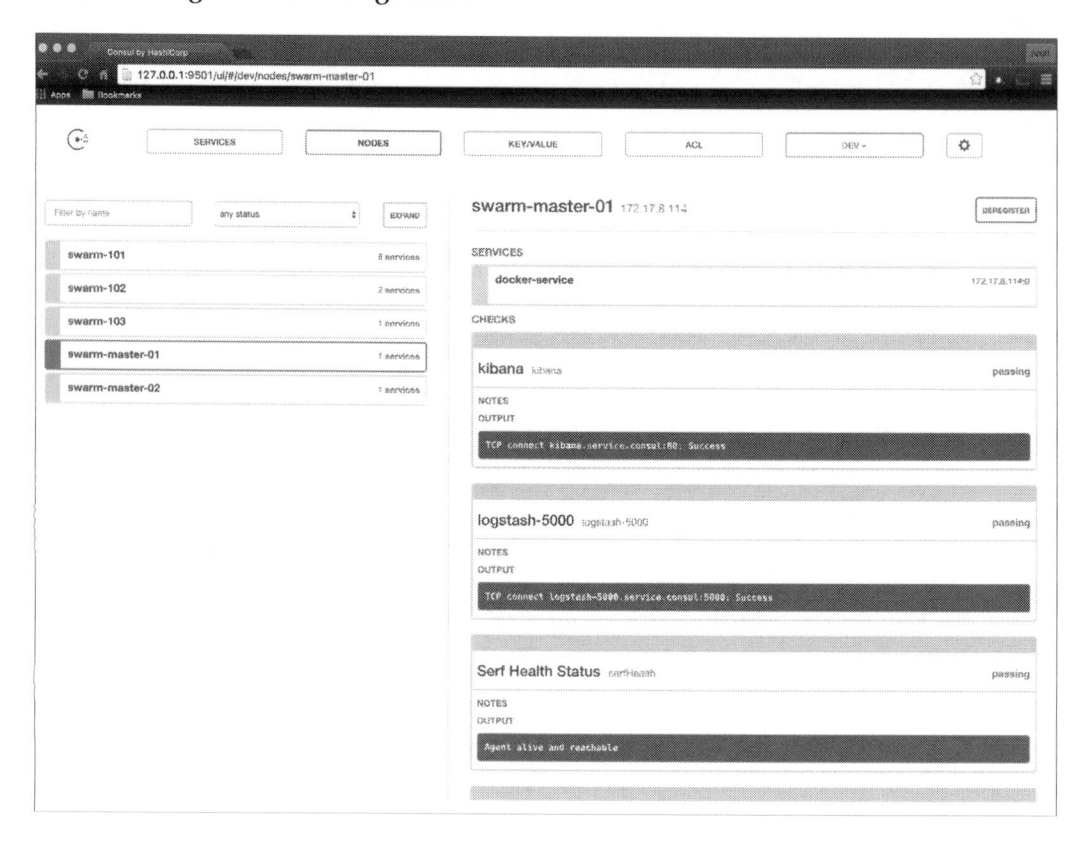

As you can see, our monitoring was successful. We will move back to our code and add a check for our swarm master to determine its health. We will do that with the following code:

[170]

```puppet
consul::service { 'swarm-master-01':
  checks  => [
    {
      script   => 'docker -H tcp://172.17.8.114:4000 info',
      interval => '10s',
      tags     => ['swarm-master-01']
    }
  ],
  address => $::ipaddress_enp0s8,
}
consul::service { 'swarm-master-02':
  checks  => [
    {
      script   => 'docker -H tcp://172.17.8.115:4000 info',
      interval => '10s',
      tags     => ['swarm-master-02']
    }
  ],
  address => $::ipaddress_enp0s8,
}
```

We will then issue the Vagrant provision command, `vagrant provision swarm-master-01 && vagrant provision swarm-master-02`. Again, we will open our web browser and click on **swarm-master-01** or **swarm-master-02**. You should get the following result after this:

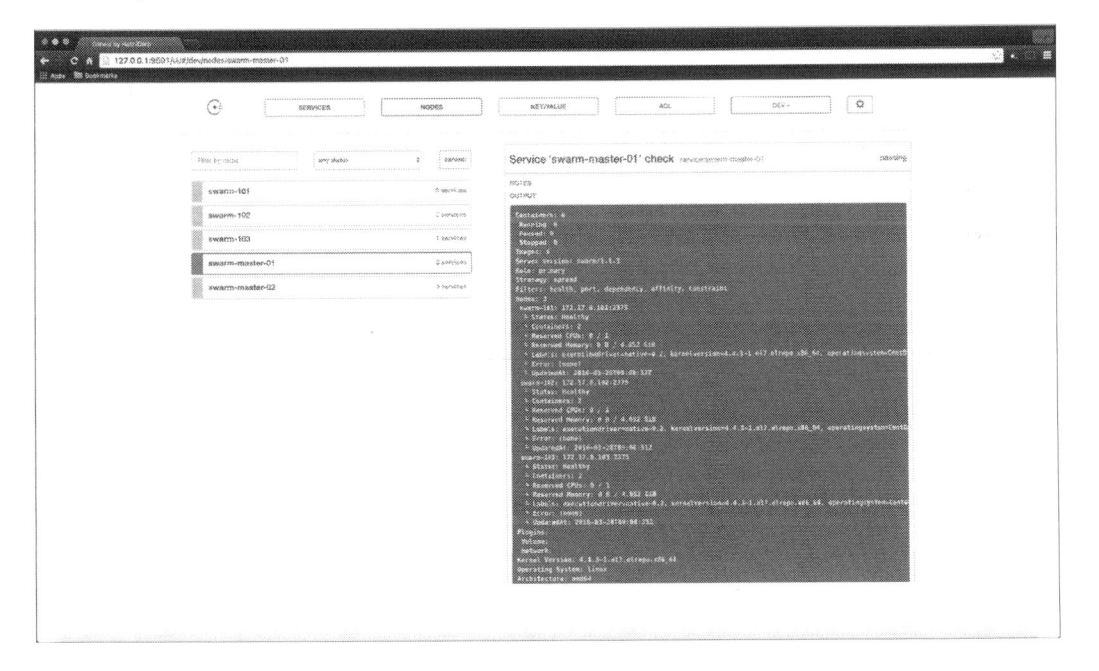

Logging, Monitoring, and Recovery Techniques

As you can see from the information in the check, we can easily see which Swarm master is primary, the strategy for scheduling. This will come in really handy when you have any issues.

So as you can see, Consul is a really handy tool, and if you want to take away the things we covered in this chapter, you can really do some cool stuff.

Recovery techniques

It is important in every solution to have some recovery techniques. In the container world, this is no different. There are many ways to skin this cat, such as load balancing with HA proxy or even using a container-based application that was built for this purpose such as interlock (https://github.com/ehazlett/interlock). If you have not checked out interlock, it's awesome!!! There are so many combinations of solutions we could use depending on the underlying application. So here, we are going to look at the built-in HA in Docker Swarm. From there, you could use something like an interlock to make sure that there is no downtime in accessing your containers.

Built-in HA

Docker Swarm has two kind of nodes: master nodes and member nodes. Each one of these has different built-in protection for failure. The first node type we will look at is master nodes.

In the last topic, we set up a health check to get the information regarding our Swarm cluster. There, we saw that we had a master or primary Swarm master and a replica. Swarm replicates all its cluster information over TCP port 4000. So, to simulate failure, we are going to turn off the master. My master is `swarm-master-01`, but yours could be different. We will use the health check that we already created to test out the failure and watch how Swarm handles itself. We will issue the `vagrant halt swarm-master-01` command. We will open up our browser again to our Consul web UI, `127.0.0.1:9501`. As we can see in the following screenshot, `swarm-master-02` is now a master:

Chapter 8

[screenshot of Consul UI showing swarm nodes]

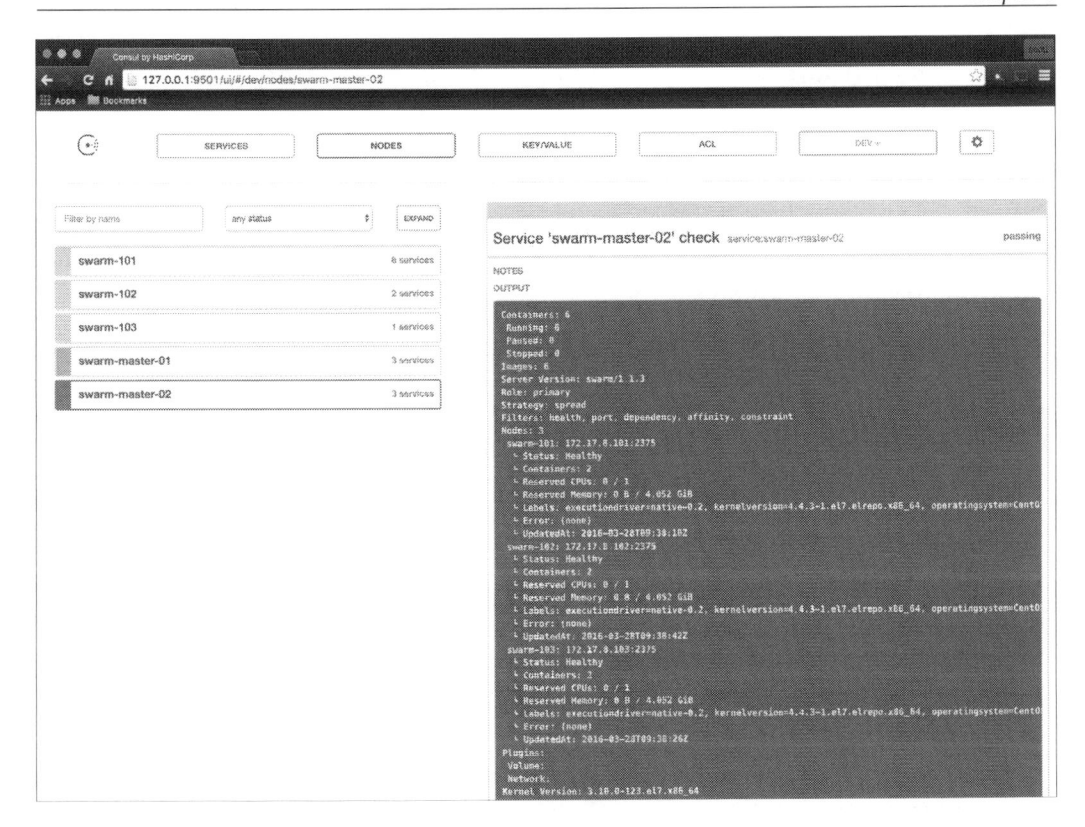

Now, we will move on to container resecluding with our Swarm node HA. As of version 1.1.3, Swarm is shipped with a feature where a container will respawn on a healthy node if the original node fails. There are some rules to this such as when you have filtering rules or linked containers. To know more on this topic, you can read the Docker docs about Swarm located at `https://github.com/docker/swarm/tree/master/experimental`.

To test this out, I will halt the node that hosts Kibana. We will need to add some code to our `kibana` container so that it will restart on failure. This is added to the `env` resource, as shown in the following screenshot:

```
swarm_run {'kibana':
    ensure      => present,
    image       => 'kibana:4.3.0',
    network     => 'swarm-private',
    ports       => ['80:5601'],
    env         => ['ELASTICSEARCH_URL=http://elasticsearch:9200', 'reschedule:on-node-failure'],
    log_driver  => 'syslog',
    log_opt     => 'syslog-address=tcp://logstash-5000.service.consul:5000',
    depends     => 'logstash',
    require     => Class['config::swarm']
}
```

Logging, Monitoring, and Recovery Techniques

We will first need to kill our old container to add the restart policy. We can do that by setting the **ensure** resource to **absent** and running the Vagrant provision `swarm-master-02`.

Once the Vagrant run is complete, we will change it back to **present** and run `vagrant provision swarm-master-02`.

For me, my `kibana` container is on **swarm-102** (this could be different for you). Once that node fails, `kibana` will restart on a healthy node. So, let's issue `vagrant halt swarm-102`. If we go to our Consul URL, `127.0.0.1:9501`, we should see some failures on our nodes and checks, as shown in the following screenshot:

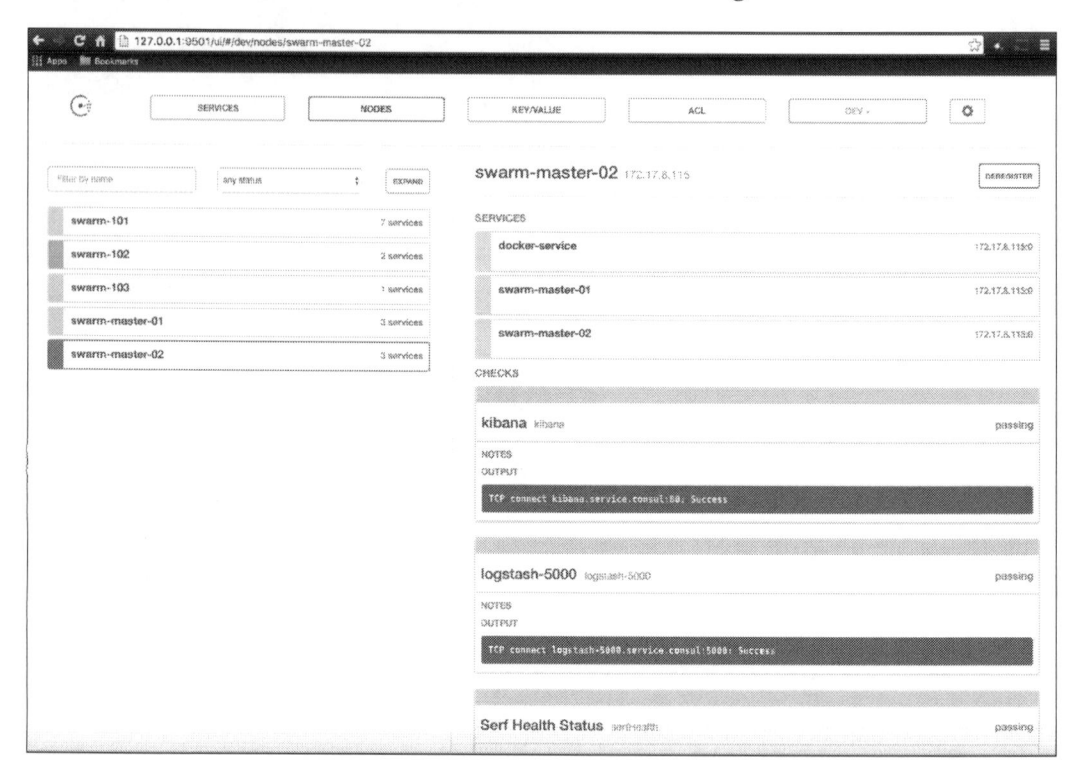

If you wait a minute or so, you will see that `kibana` alerts came back and the container spawned on another server. For me, `kibana` came back on **swarm-101**, as you can see in the following screenshot:

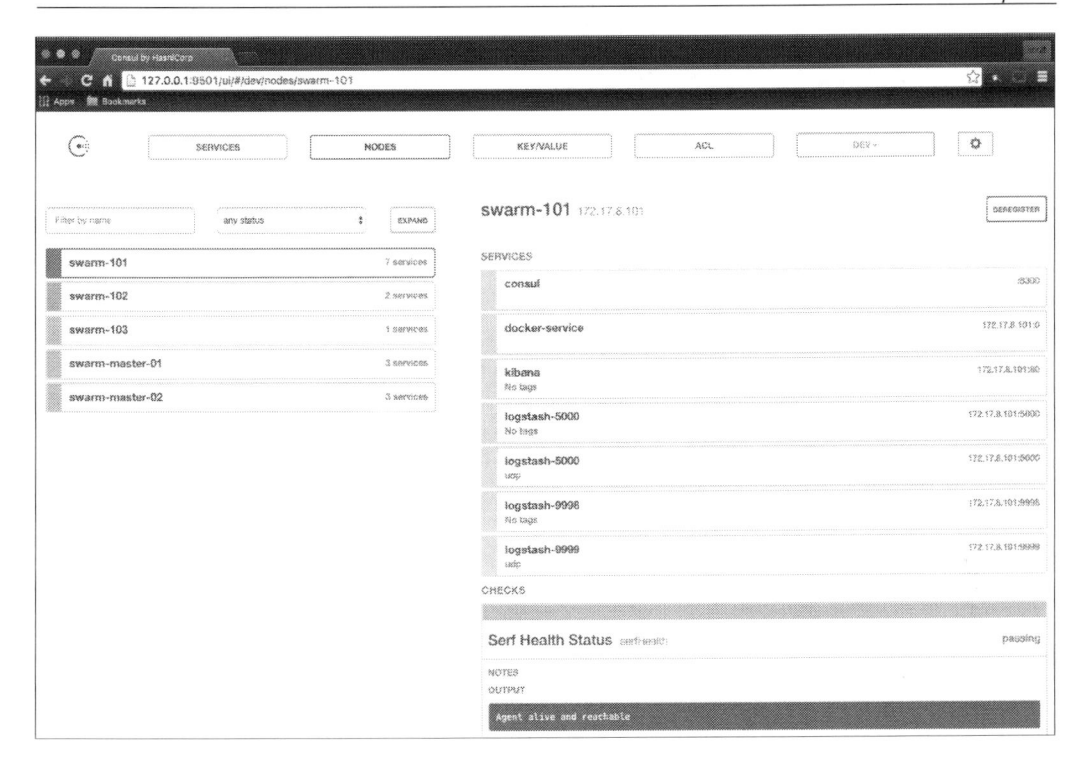

We can then go to our browser and look at `kibana`. For me, it will be at
`127.0.0.1:8001`:

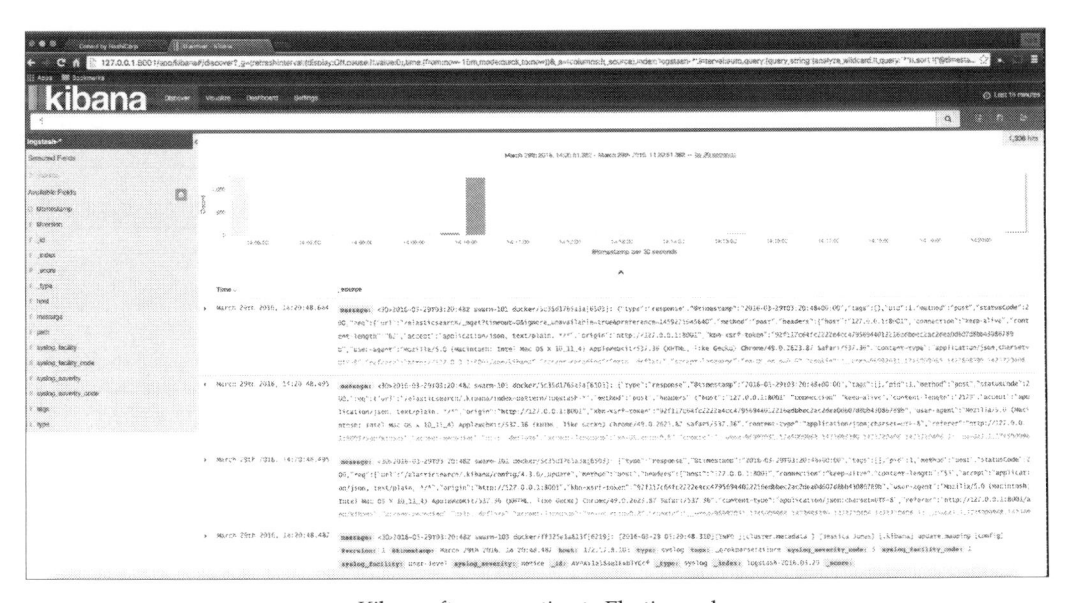

Kibana after connecting to Elasticsearch

As you can see, all our logs are there; our service discovery worked perfectly as once the container changed nodes, our health checks turned green.

Summary

In this chapter, we looked at how to operationalize our container environment using Puppet. We covered a logging solution using ELK. We took Consul to the next level with more in-depth health checks and creating services to monitor our cluster. We then tested the built-in HA functionality that ships with Swarm. We have covered a lot of ground now since our humble first module in *Chapter 2, Working with Docker Hub*. You are fully equipped to take the knowledge that you have got here and apply it in the real world.

9
Best Practices for the Real World

We have really covered a lot of ground in this book so far. We are now on our final chapter. Here, we are going to look at how to combine all the skills you have learned and create a production-ready module. Just to go out with a bang, we will create a module that configures and deploys Kubernetes as the frontend (note that running Kuberbetes as frontend has limitations and would not be the best idea for production. The UCP component of the module will be production ready). Since there will be a lot of sensitive data, we will take advantage of Hiera. We will create a custom fact to automate the retrieval of the UCP fingerprint, and we will split out all the kubernetes components and use interlock to proxy our API service. We will also take UCP further and look at how to repoint the Docker daemon to use the UCP cluster. The server architecture will follow the same design as we discussed in the scheduler chapter. We will use three nodes, all running Ubuntu 14.04 with an updated kernel to support the native Docker network namespace. We will cover the following topics in this chapter:

- Hiera
- The code

Hiera

In this topic, we will look at how to make our modules stateless. We will move all the data that is specific to the node that is applied to the module in Hiera (https://docs.puppetlabs.com/hiera/3.1/). There are two main drivers behind this. The first is to remove any sensitive data such as passwords, keys, and so on, out of our modules. The second is if we remove node-specific data or state out our modules so that they are generic. We can apply them to any number of hosts without changing the logic of the module. This gives us the flexibility to publish our modules for other members of the Puppet community.

What data belongs in Hiera

When we first sit down and start development on a new module, some of the things that we should consider are: whether we can make our module OS agnostic, how we can run the module on multiple machines without logic changes or extra development, and how we can protect our sensitive data?

The answer to all these questions is Hiera.

We are able to leverage Hiera by parameterizing our classes. This will allow Puppet to automatically look up Hiera at the beginning of the catalogue compilation. So, let's explore some examples of the data that you will put in Hiera.

Note that in our chapter about container schedulers, we briefly used Hiera. We set the following values:

```
1  docker_swarm::swarm_version: v1.1.3
2  docker_swarm::backend: consul
3  docker_swarm::backend_ip: 172.17.8.101
4  docker_swarm::backend_port: 8500
5  docker_swarm::advertise_int: enp0s8
6  consul::version: 0.6.3
```

As you can see, we are setting parameters such as versions. Why would we want to set versions in Hiera and not straight in the module? If we set the version of Consul, for example, we might be running version 5.0 in production. Hashicorp just released version 6.0 by changing the Hiera value for the environment (for more information about Puppet environments, go to https://docs.puppetlabs.com/puppet/latest/reference/environments.html). In Hiera versions from dev to 6.0, we can run multiple versions of the application with no module development.

The same can be done with IP addresses or URLs. So in the hieradata for dev, your Swarm cluster URL could be dev.swarm.local and production could be just swarm.local.

Another type of data that you will want to separate is passwords/keys. You wouldn't want the same password/key in your dev environment as there are in production. Again, Hiera will let you obfuscate this data.

We can then take the protection of this data further using eyaml, which Puppet supports. This allows you to use keys to encrypt your Hiera .yaml files. So, when the files are checked into source control, they are encrypted. This helps prevent data leakage. For more information on eyaml, visit https://puppetlabs.com/blog/encrypt-your-data-using-hiera-eyaml.

As you see, Hiera gives you the flexibility to move your data from the module to Hiera to externalize configurations, making the module stateless.

Tips and tricks for Hiera

There are some really handy tips that you can refer to when using Hiera. Puppet allows you to call functions, lookups, and query facts from Hiera. Mastering these features will come in very handy. If this is new to you, read the document available at https://docs.puppetlabs.com/hiera/3.1/variables.html before moving on in the chapter.

So, let's look at an example of looking up facts from Hiera. First, why would you want to do this? One really good reason is IP address lookups. If you have a class that you are applying to three nodes and you have to advertise an IP address like we did in our `consul` module, setting the IP address in Hiera will not work, as the IP address will be different for each machine. We create a file called `node.yaml` and add the IP address there. The issue is that we will now have multiple Hiera files. Every time Puppet loads the catalogue, it looks up all the Hiera files to check whether any values have changed. The more files we have, the more load it puts on the master and the slower our Puppet runs will become. So we can tell Hiera to look up the fact and the interface we want to advertise. Here is an example of what the code would look like:

```
ucpconfig::ucp_host_address: "%{::ipaddress_eth1}"
```

The one call out is that if we called the fact from a module, we would call it with the fully qualified name, `$::ipaddress_eth1`. Unfortunately, Hiera does not support the use of this. So we can use the short name for the `::ipaddress_eth1` fact.

The code

Now that we have a good understanding of how to make our module stateless, let's go ahead and start coding. We will split the coding into two parts. In the first part we will write the module to install and configure Docker UCP. The final topic will be to run Kubernetes as the frontend.

UCP

The first thing that we need to do is create a new Vagrant repo for this chapter. By now, we should be masters at creating a new Vagrant repo. Once we have created that, we will create a new module called `<AUTHOR>-ucpconfig` and move it to our `modules` directory in the root of our Vagrant repo. We will first set up our `servers.yml` file by adding the code shown in the following screenshot:

```yaml
---
-
  box: scottyc/ubuntu-14-04-puppet-kernel-4-2
  cpu: 2
  ip: "172.17.10.101"
  name: ucp-01
  forward_ports:
    - { guest: 443, host: 8443 }
  ram: 4096
  shell_commands:
    - { shell: 'apt-get update -y' }
    - { shell: 'apt-get install -y wget git' }
    - { shell: 'mkdir ~/.docker || true && cp /vagrant/config.json ~/.docker/' }
    - { shell: 'mkdir /etc/docker/ || true && cp /vagrant/docker_subscription.lic /etc/docker/subscription.lic' }
    - { shell: '/opt/puppet/bin/gem install r10k && ln -s /opt/puppet/bin/r10k /usr/bin/r10k || true'}
    - { shell: 'cp /home/vagrant/ucp-01/Puppetfile /tmp && cd /tmp && r10k puppetfile install --verbose' }
    - { shell: 'cp /home/vagrant/ucp-01/modules/* -R /tmp/modules || true' }

-
  box: scottyc/ubuntu-14-04-puppet-kernel-4-2
  cpu: 2
  ip: "172.17.10.102"
  name: ucp-02
  forward_ports:
    - { guest: 443, host: 9443 }
  ram: 4096
  shell_commands:
    - { shell: 'apt-get update -y' }
    - { shell: 'apt-get install -y wget git' }
    - { shell: 'mkdir ~/.docker || true && cp /vagrant/config.json ~/.docker/' }
    - { shell: 'mkdir /etc/docker/ || true && cp /vagrant/docker_subscription.lic /etc/docker/subscription.lic' }
    - { shell: '/opt/puppet/bin/gem install r10k && ln -s /opt/puppet/bin/r10k /usr/bin/r10k || true'}
    - { shell: 'cp /home/vagrant/ucp-02/Puppetfile /tmp && cd /tmp && r10k puppetfile install --verbose' }
    - { shell: 'cp /home/vagrant/ucp-02/modules/* -R /tmp/modules || true' }

-
  box: scottyc/ubuntu-14-04-puppet-kernel-4-2
  cpu: 2
  ip: "172.17.10.103"
  name: ucp-03
  forward_ports:
    - { guest: 443, host: 10443 }
  ram: 4096
  shell_commands:
    - { shell: 'apt-get update -y' }
    - { shell: 'apt-get install -y wget git' }
    - { shell: 'mkdir ~/.docker || true && cp /vagrant/config.json ~/.docker/' }
    - { shell: 'mkdir /etc/docker/ || true && cp /vagrant/docker_subscription.lic /etc/docker/subscription.lic' }
    - { shell: '/opt/puppet/bin/gem install r10k && ln -s /opt/puppet/bin/r10k /usr/bin/r10k || true'}
    - { shell: 'cp /home/vagrant/ucp-03/Puppetfile /tmp && cd /tmp && r10k puppetfile install --verbose' }
    - { shell: 'cp /home/vagrant/ucp-03/modules/* -R /tmp/modules || true' }
```

As you can see, we are setting up three servers, where `ucp-01` will be the master of the cluster and the other two nodes will join the cluster. We will add two more files: `config.json` and `docker_subscription.lic`. Here, `config.json` will contain the authorization key to Docker Hub, and `docker_subscription.lic` will contain our trial license for UCP. Note that we covered both of these files in the container scheduler chapter. If you are having issues setting up these files, refer to that chapter.

The file we will now look at is the puppetfile. We need to add the code shown in the following screenshot to it:

```ruby
#!/usr/bin/ruby env

require "socket"
$hostname = Socket.gethostname

forge 'http://forge.puppetlabs.com'

mod 'garethr/docker', :git => 'https://github.com/scotty-c/garethr-docker.git'
mod 'puppetlabs/apt'
mod 'puppetlabs/docker_ucp'
mod 'puppetlabs/stdlib'
mod 'maestrodev/wget'
```

Now that we have our Vagrant repo set up, we can move on to our module. We will need to create four files: `config.pp`, `master.pp`, `node.pp`, and `params.pp`.

The first file we will look at is `params.pp`. Again, I'd like to code this file first as it sets a good foundation for the rest of the module. We do this as follows:

```puppet
class ucpconfig::params {
  $ucp_master                     = ''
  $ucp_deploy_node                = ''
  $ucp_url                        = ''
  $ucp_username                   = ''
  $ucp_password                   = ''
  $ucp_fingerprint                = $::ucp_fingerprint
  $ucp_version                    = '1.0.0'
  $ucp_host_address               = ''
  $ucp_subject_alternative_names  = ''
  $ucp_external_ca                = false
  $ucp_swarm_scheduler            = 'binpack'
  $ucp_swarm_port                 = ''
  $ucp_controller_port            = '8443'
  $ucp_preserve_certs             = 'true'
  $ucp_license_file               = ''
  $consul_master_ip               = ''
  $consul_advertise               = ''
  $consul_image                   = 'scottyc/consul'
  $consul_bootstrap_num           = '1'
  $docker_network                 = 'private-net'
  $docker_network_driver          = 'overlay'
  $docker_cert_path               = ''
  $docker_host                    = ''
}
```

Best Practices for the Real World

As you can see, we are setting all our parameters. We will look at each one in depth as we apply it to the class. This way, we will have context on the value we are setting. You might have noted that we have set a lot of variables to empty strings. This is because we will use Hiera to look up the values. I have hardcoded some values in as well, such as the UCP version as 1.0.0. This is just a default value if there is no value in Hiera. However, we will be setting the value for UCP to 1.0.3. The expected behavior is that the Hiera value will take precedence. You will note that there is a fact that we are referencing, which is $::ucp_fingerprint. This is a custom fact. This will automate the passing of the UCP fingerprint. If you remember, in the container scheduler chapter, we had to build `ucp-01` to get the fingerprint and add it to Hiera for the other nodes' benefit. With the custom fact, we will automate the process.

To create a custom fact, we will first need to create a `lib` folder in the root of the module. Under that folder, we will create a folder called `facter`. In that folder, we will create a file called `ucp_fingerprint.rb`. When writing a custom fact, the filename needs to be the same as the fact's name. The code that we will add to our custom fact is as follows:

```
Facter.add('ucp_fingerprint') do
  setcode do
    Facter::Core::Execution.exec("echo -n | openssl s_client -connect 172.17.10.101:443 2> /dev/null | sed -ne '/-BEGIN CERTIFICATE-/,/-END CERTIFICATE-/p' | openssl x509 -noout -fingerprint -sha1 | cut -d= -f2")
  end
end
```

In this code, you can see that we are adding a `bash` command to query our UCP master in order to get the fingerprint. I have hardcoded the IP address of the master in this fact. In our environment, we would write logic for that to be more fluid so that we are allowed to have multiple environments, hostnames, and so on. The main part to take away from the custom fact is the command itself.

We will now move back to our `init.pp` file, which is as follows:

```puppet
class ucpconfig {

  $ucp_master                    = $ucpconfig::params::ucp_master,
  $ucp_deploy_node               = $ucpconfig::params::ucp_deploy_node,
  $ucp_url                       = $ucpconfig::params::ucp_url,
  $ucp_username                  = $ucpconfig::params::ucp_username,
  $ucp_password                  = $ucpconfig::params::ucp_password,
  $ucp_fingerprint               = $ucpconfig::params::ucp_fingerprint,
  $ucp_version                   = $ucpconfig::params::ucp_version,
  $ucp_host_address              = $ucpconfig::params::ucp_host_address,
  $ucp_subject_alternative_names = $ucpconfig::params::ucp_subject_alternative_names,
  $ucp_external_ca               = $ucpconfig::params::ucp_external_ca,
  $ucp_swarm_scheduler           = $ucpconfig::params::ucp_swarm_scheduler,
  $ucp_swarm_port                = $ucpconfig::params::ucp_swarm_port,
  $ucp_controller_port           = $ucpconfig::params::ucp_controller_port,
  $ucp_preserve_certs            = $ucpconfig::params::ucp_preserve_certs,
  $ucp_license_file              = $ucpconfig::params::ucp_license_file,
  $consul_master_ip              = $ucpconfig::params::consul_master_ip,
  $consul_advertise              = $ucpconfig::params::consul_advertise,
  $consul_image                  = $ucpconfig::params::consul_image,
  $consul_bootstrap_num          = $ucpconfig::params::consul_bootstrap_num,
  $docker_network                = $ucpconfig::params::docker_network,
  $docker_network_driver         = $ucpconfig::params::docker_network_driver,
  $docker_cert_path              = $ucpconfig::params::docker_cert_path,
  $docker_host                   = $ucpconfig::params::docker_host,
) inherits ucpconfig::params {

  class { 'docker':
    tcp_bind         => 'tcp://127.0.0.1:4243',
    socket_bind      => 'unix:///var/run/docker.sock',
    extra_parameters => "--cluster-store=consul://${consul_master_ip}:8500 --cluster-advertise=${consul_advertise}",
  } ->

  case $::hostname {
    "$ucp_master": {

      docker::image { $consul_image: } ->

      docker::run { 'consul':
        image    => $consul_image,
        hostname => 'consul',
        command  => "-server --advertise ${consul_master_ip} -bootstrap-expect ${consul_bootstrap_num}",
        ports    => ['8301:8301', '8301:8301/udp', '8302:8302', '8302:8302/udp', '8400:8400', '8500:8500', '8600:8600', '8600:8600/udp'],
        before   => Class['ucpconfig::master']
      }
      contain ucpconfig::master
      contain ucpconfig::config


      Class['ucpconfig::master'] -> Class['ucpconfig::config']

    }
    "$ucp_deploy_node": {
      include ucpconfig::node
      contain ucpconfig::config
      contain ucpconfig::compose

      Class['ucpconfig::config'] -> Class['ucpconfig::node'] -> Class['ucpconfig::compose']
    }
    default: {
      include ucpconfig::node
      contain ucpconfig::config


      Class['ucpconfig::config'] -> Class['ucpconfig::node']
    }
  }
}
```

The first thing you can see it that we are declaring all our variables at the top of our class. This is where Puppet will look up Hiera and match any variable that we have declared. The first block of code in the module is going to set up our Docker daemon. We are going to add the extra configuration to tell the daemon where it can find the backend for the Docker native network. Then, we will declare a case statement on the `$::hostname` fact. You will note that we have set a parameter for the hostname. This is to make our module more portable. Inside the case statement, if the host is a master, you can see that we will use our Consul container as the backend for our Docker network. Then, we will order the execution of our classes that are applied to the node.

In the next block of code in the case statement, we have declared the `$ucp_deploy_node` variable for the `$::hostname` fact. We will use this node to deploy Kubernetes from. We will get back to this later in the topic. The final block of code in our case statement is the `catch all` or `default`. If Puppet cannot find the fact of `$::hostname` in either of our declared variables, it will apply these classes.

We will now move on to our `master.pp` file, which is as follows:

```
class ucpconfig::master(

  $ucp_version                      = $ucpconfig::ucp_version,
  $ucp_host_address                 = $ucpconfig::ucp_host_address,
  $ucp_subject_alternative_names    = $ucpconfig::ucp_subject_alternative_names,
  $ucp_external_ca                  = $ucpconfig::ucp_external_ca,
  $ucp_swarm_scheduler              = $ucpconfig::ucp_swarm_scheduler,
  $ucp_swarm_port                   = $ucpconfig::ucp_swarm_port,
  $ucp_controller_port              = $ucpconfig::ucp_controller_port,
  $ucp_preserve_certs               = $ucpconfig::ucp_preserve_certs,
  $ucp_license_file                 = $ucpconfig::ucp_license_file,

) {

  class { 'docker_ucp':
    controller                      => true,
    host_address                    => $ucp_host_address,
    version                         => $ucp_version,
    usage                           => false,
    tracking                        => false,
    subject_alternative_names       => $ucp_subject_alternative_names,
    external_ca                     => $ucp_external_ca,
    swarm_scheduler                 => $ucp_swarm_scheduler,
    swarm_port                      => $ucp_swarm_port,
    controller_port                 => $ucp_controller_port,
    preserve_certs                  => $ucp_preserve_certs,
    docker_socket_path              => '/var/run/docker.sock',
    license_file                    => $ucp_license_file,
    require                         => Class['docker']
  }
}
```

The first thing you will note is that we are declaring our variables again at the top of this class. We are doing this as we are declaring many parameters, and this makes our module a lot more readable. Doing this in complex modules is a must, and I would really recommend you to follow this practice. It will make your life a lot easier when it comes to debugging. As you can see in the following screenshot, we are tying the parameters back to our `init.pp` file:

```puppet
class { 'docker_ucp':
    controller                => true,
    host_address              => $ucp_host_address,
    version                   => $ucp_version,
    usage                     => false,
    tracking                  => false,
    subject_alternative_names => $ucp_subject_alternative_names,
    external_ca               => $ucp_external_ca,
    swarm_scheduler           => $ucp_swarm_scheduler,
    swarm_port                => $ucp_swarm_port,
    controller_port           => $ucp_controller_port,
    preserve_certs            => $ucp_preserve_certs,
    docker_socket_path        => '/var/run/docker.sock',
    license_file              => $ucp_license_file,
    require                   => Class['docker']
}
```

As you can see from the preceding code, there are very few types which we are values that we are setting in the module.

We will now move on to our `node.pp` file, as follows:

```puppet
class ucpconfig::node (

    $ucp_url                       = $ucpconfig::ucp_url,
    $ucp_username                  = $ucpconfig::ucp_username,
    $ucp_password                  = $ucpconfig::ucp_password,
    $ucp_fingerprint               = $ucpconfig::ucp_fingerprint,
    $ucp_version                   = $ucpconfig::ucp_version,
    $ucp_host_address              = $ucpconfig::ucp_host_address,
    $ucp_subject_alternative_names = $ucpconfig::ucp_subject_alternative_names,

){

    class { 'docker_ucp':
        ucp_url                   => $ucp_url,
        fingerprint               => $ucp_fingerprint,
        username                  => $ucp_username,
        password                  => $ucp_password ,
        host_address              => $ucp_host_address,
        subject_alternative_names => $ucp_subject_alternative_names,
        replica                   => true,
        version                   => $ucp_version,
        usage                     => false,
        tracking                  => false,
        require                   => Class['docker']
    }
}
```

Best Practices for the Real World

As you can see, we are declaring the parameters at the top of the class, again tying them back to our `init.pp` file. We have declared most of our values as we will use Hiera.

We will now move on to our `config.pp` file, as follows:

```puppet
class ucpconfig::config (

  $ucp_url              = $ucpconfig::ucp_url,
  $ucp_username         = $ucpconfig::ucp_username,
  $ucp_password         = $ucpconfig::ucp_password,
  $docker_network       = $ucpconfig::docker_network,
  $docker_network_driver = $ucpconfig::docker_network_driver,
  $docker_cert_path     = $ucpconfig::docker_cert_path,
  $docker_host          = $ucpconfig::docker_host,
) {


  package { ['curl', 'zip', 'jq']:
    ensure => installed,
  }

  file { '/etc/docker/get_ca.sh':
    ensure  => file,
    content => template("ucpconfig/get_ca.sh.erb"),
  }

  exec { 'ca_bundle':
    command => 'sh get_ca.sh',
    path    => '/usr/bin:/usr/sbin:/bin:/usr/local/bin',
    cwd     => $docker_cert_path,
    creates => "${docker_cert_path}/bundle.zip",
    require => File['/etc/docker/get_ca.sh']
  }

  file { '/etc/profile.d/docker.sh':
    ensure  => present,
    content => template('ucpconfig/docker.sh.erb'),
    mode    => '0644',
  }


  docker_network { $docker_network:
    ensure  => present,
    create  => true,
    driver  => $docker_network_driver,
    require => File['/etc/profile.d/docker.sh']
  }
}
```

In this class, we are going to make a few vital configurations for our cluster. So, we will walk through each block of code individually. Let's take a look at the first one:

```
class ucpconfig::config (
    $ucp_url                = $ucpconfig::ucp_url,
    $ucp_username           = $ucpconfig::ucp_username,
    $ucp_password           = $ucpconfig::ucp_password,
    $docker_network         = $ucpconfig::docker_network,
    $docker_network_driver  = $ucpconfig::docker_network_driver,
    $docker_cert_path       = $ucpconfig::docker_cert_path,
    $docker_host            = $ucpconfig::docker_host,
) {
```

In this block, we will declare our variables as we have in all the classes in this module. One call out that I have not mentioned yet is that we are only declaring the variables that are applied to its class. We are not declaring all the parameters. Let's look at the next set of code now:

```
package { ['curl', 'zip', 'jq']:
    ensure => installed,
}
```

In this block of code, we will pass an array of packages that we need to curl our UCP master in order to get the SSL bundle that we will need for TLS coms between nodes. Now, let's see the third block:

```
file { '/etc/docker/get_ca.sh':
    ensure  => file,
    content => template("ucpconfig/get_ca.sh.erb"),
}
```

In this code, we are going to create a shell script called `get_ca.sh.erb` to get the bundle from the master. Let's create the file, and the first thing that we will need to do is create a `templates` folder in the root of the module. Then, we can create our `get_ca.sh.erb` file in the `templates` folder. We will add the following code to the file:

```
1  #!/bin/bash
2  AUTHTOKEN=$(curl -sk -d '{"username":"<%= @ucp_username %>","password":"<%= @ucp_password %>"}' <%= @ucp_url %>/auth/login | jq -r .auth_token) && \
3  curl -k -H "Authorization: Bearer $AUTHTOKEN" <%= @ucp_url %>/api/clientbundle -o bundle.zip && \
4  unzip bundle.zip
```

As you can see in the script, we need to create an `auth` token and pass it to the API. Puppet does not natively handle tasks like these well, as we are using variables inside the `curl` command. Creating a template file and running an `exec` function is fine as long as we make it idempotent. In the next block of code, we will do this:

```
exec { 'ca_bundle':
    command => 'sh get_ca.sh',
    path    => '/usr/bin:/usr/sbin:/bin:/usr/local/bin',
    cwd     => $docker_cert_path,
    creates => "${$docker_cert_path}/ca.pem",
    require => File['/etc/docker/get_ca.sh']
}
```

In this block of code, we will run the preceding script. You can see that we have set the parameters, `command path`, and **cwd** (**current working directory**). The next resource is `creates` that tells Puppet that there should be a file called `ca.pem` in the current working directory. If Puppet finds that the file does not exist, it will execute the exec, and if the file does exist, Puppet will do nothing. This will give our exec idempotency.

In the next block of code, we will create a file in `/etc/profile.d`, which will then point the Docker daemon on each node to the master's IP address, allowing us to schedule containers across the cluster:

```
file { '/etc/profile.d/docker.sh':
    ensure  => present,
    content => template('ucpconfig/docker.sh.erb'),
    mode    => '0644',
}
```

Now, let's create a `docker.sh` file in our `templates` directory. In the file, we will put the following code:

```
#!/bin/bash
export DOCKER_TLS_VERIFY=1
export DOCKER_CERT_PATH=<%= @docker_cert_path %>
export DOCKER_HOST=<%= @docker_host %>
```

In this file, we are telling the Docker daemon to use TLS, setting the location of the key files that we got from the bundle earlier in the class. The last thing we are setting is the Docker host that will point to the UCP master.

This last block of code in this class should look very familiar, as we are setting up a Docker network:

```
docker_network { $docker_network:
    ensure  => present,
    create  => true,
    driver  => $docker_network_driver,
    require => File['/etc/profile.d/docker.sh']
    }
}
```

We can now move on to the the file where all our data is, our Hiera file. That file is located in the `hieradata` folder, which is present in the root of our Vagrant repo. The following screenshot shows the various data present in the Hiera file:

```
1   ucpconfig::ucp_master: ucp-01
2   ucpconfig::ucp_deploy_node: ucp-03
3   ucpconfig::ucp_url: https://172.17.10.101
4   ucpconfig::ucp_username: admin
5   ucpconfig::ucp_password: orca
6   ucpconfig::ucp_version: 1.0.3
7   ucpconfig::ucp_host_address: "%{::ipaddress_eth1}"
8   ucpconfig::ucp_subject_alternative_names: "%{::ipaddress_eth0}"
9   ucpconfig::ucp_external_ca: false
10  ucpconfig::ucp_swarm_scheduler: spread
11  ucpconfig::ucp_swarm_port: 19001
12  ucpconfig::ucp_controller_port: 443
13  ucpconfig::ucp_preserve_certs: true
14  ucpconfig::ucp_license_file: /etc/docker/subscription.lic
15  ucpconfig::consul_master_ip: 172.17.10.101
16  ucpconfig::consul_advertise: eth1:2376
17  ucpconfig::consul_image: scottyc/consul
18  ucpconfig::consul_bootstrap_num: 1
19  ucpconfig::docker_network: swarm-private
20  ucpconfig::docker_network_drive: overlay
21  ucpconfig::docker_cert_path: /etc/docker
22  ucpconfig::docker_host: tcp://172.17.10.101:443
23
```

So, let's list out all the data we have defined here. We are defining the UCP master as `ucp-01`, and our deploy node is `ucp-03` (this is the node that we are deploying Kubernetes from). The UCP URL of the master is `https://172.17.10.101`. This is used when we connect nodes to the master and also when we get our `ca` bundle and UCP fingerprint. We keep the username and password as `admin` and `orca`. We will use UCP version `1.0.3`. We will then use the Hiera fact lookup that we discussed earlier in the book to set the host address and alternate names for UCP.

Best Practices for the Real World

The next parameter will tell UCP that we will use an internal CA. We will then set our scheduler to use `spread`, define the ports for both Swarm and the controller, tell UCP to persevere the certs, and send the location of the license file for UCP. Next, we will set some date for Consul, such as the master IP, the interface to advertise, the image to use, and how many nodes to expect at the time of booting the Consul cluster. Lastly, we will set the variables we want our Docker daemon to use, such as the name of our Docker network, `swarm-private`, the network driver, `overlay`, the path to our certs, and the Docker host that we set in our `docker.sh` file placed at `/etc/profile.d/`.

So, as you can see, we have a whole lot of data in Hiera. However, as we have already discussed, there is data that could be changed for different environments. So as you can see, there is a massive benefit in making your module stateless and abstracting your data to Hiera, especially if you want to write modules that can scale easily.

We will now add the following code to our manifest file, `default.pp` located in the `manifests` folder in the root of our Vagrant repo. The following code defines our node definition:

```
1  node 'ucp-01' {
2    include ucpconfig
3  }
4
5  node 'ucp-02' {
6    include ucpconfig
7  }
8
9  node 'ucp-03' {
10   include ucpconfig
11 }
12
13
```

We can then open our terminal and change the directory to the root of our Vagrant repo. We will then issue the `vagrant up` command to run Vagrant. Once the three boxes are built, you should get the following terminal output:

We can then log in to the web URL at `https://127.0.0.1:8443`:

We will log in with the `admin` username and `orca` password. In the following screenshot, we can see that our cluster is up and healthy:

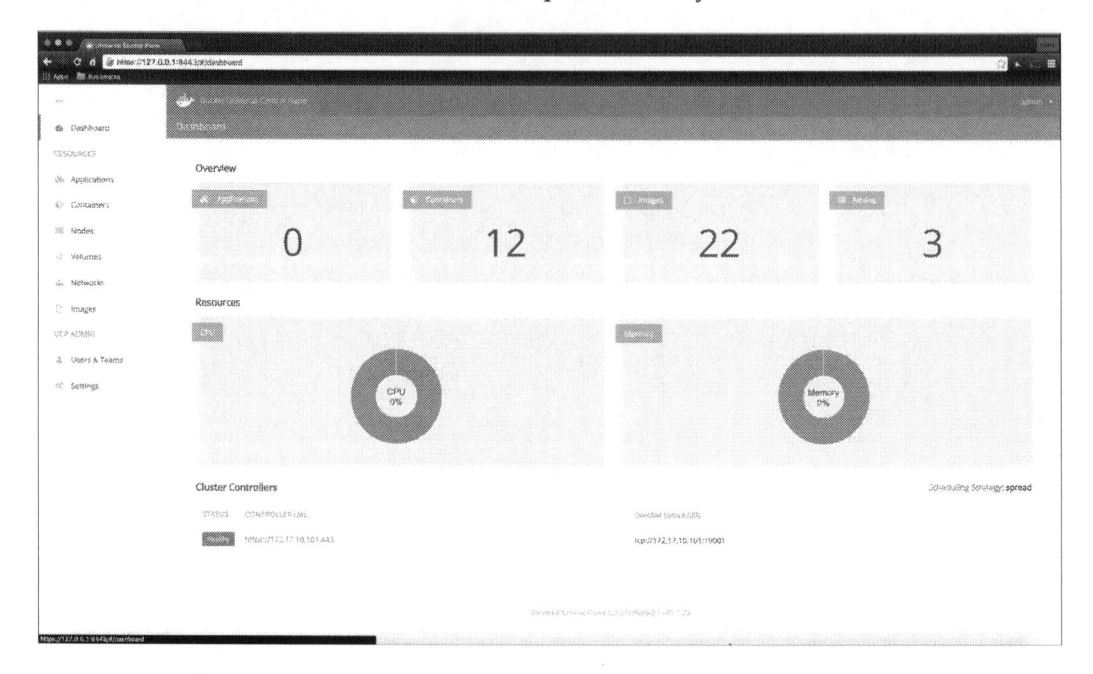

Kubernetes

Since I thought to finish off with a bang, we will do something pretty cool now. I wanted to deploy an application that had multiple containers in which we could use interlock to showcase application routing/load balancing. What better application than Kubernetes. As I mentioned earlier, there are limitations to running Kubernetes like this, and it is only for lab purposes. The skills we can take away and apply to our Puppet modules is application routing/load balancing. In our last topic, we set the parameter for `$ucp_deploy_node` in the case statement in our `init.pp` file. In that particular block of code, we had a class called `compose.pp`. This is the class that will deploy Kubernetes across our UCP cluster. Let's look at the file:

```puppet
class ucpconfig::compose {

  file { '/etc/kubernetes':
    ensure => directory,
  }

  file { '/etc/kubernetes/docker-compose.yml':
    ensure  => file,
    content => template('ucpconfig/kubernertes.yml.erb'),
    require => File['/etc/kubernetes'],
  }

  exec { 'docker-compose':
    command => 'bash -l -c "docker-compose -f /etc/kubernetes/docker-compose.yml up -d"',
    path    => '/usr/bin:/usr/sbin:/bin:/usr/local/bin',
    unless  => 'docker inspect -f {{.State.Running}} apiserver',
  }
}
```

The first resource just creates a directory called kubernetes. This is where we will place our Docker Compose file. Now, you will notice something different about how we are running Docker Compose. An exec? Why would we use an exec when we have a perfectly good provider that is tried and trusted. The reason we are using the exec in this case is because we are changing $PATH during the last Puppet run. What do I mean by that? Remember the file we added to the /etc/profile.d/ directory? It changed the shell settings for where DOCKER_HOST is pointing. This will only come into effect in the next Puppet run, so the Docker daemon will not be pointing to the cluster. This will mean that all the kubernetes containers will come up on one host. This will cause a failure in the catalogue, as we will get a port collision from two containers using 8080. Now, this will only come into effect when we run the module all at once, as a single catalogue.

Best Practices for the Real World

Now, let's have a look at our Docker Compose file:

```yaml
version: '2'
services:
  interlock:
    image: ehazlett/interlock
    container_name: interlock
    ports:
      - "80:80"
      - "8080:8080"
      - "8443:8443"
    volumes:
      - /etc/docker:/etc/docker
    command: "--swarm-url tcp://172.17.10.101:443 --swarm-tls-ca-cert /etc/docker/ca.pem --swarm-tls-cert /etc/docker/cert.pem --swarm-tls-key /etc/docker/key.pem --plugin haproxy start"
    network_mode: swarm-private
    environment:
      - "constraint:node==ucp-03"
  etcd:
    image: gcr.io/google_containers/etcd:2.2.1
    container_name: etcd
    network_mode: swarm-private
    command: ['/usr/local/bin/etcd', '--bind-addr=0.0.0.0:4001', '--data-dir=/var/etcd/data']
  apiserver:
    image: gcr.io/google_containers/hyperkube:v1.1.8
    container_name: apiserver
    ports:
      - "8080:8080"
    network_mode: swarm-private
    command: ["/hyperkube", "apiserver", "--service-cluster-ip-range=10.17.17.1/24", "--address=0.0.0.0", "--etcd_servers=http://etcd:4001", "--cluster_name=kubernetes", "--v=2"]
    environment:
      - INTERLOCK_DATA={"hostname":"kubernetes","domain":"ucp-demo.local"}
  controller:
    image: gcr.io/google_containers/hyperkube:v1.1.8
    command: ["/hyperkube", "controller-manager", "--address=0.0.0.0", "--master=http://apiserver:8080", "--v=2"]
    network_mode: swarm-private
    environment:
      - "affinity:container==apiserver"
  scheduler:
    image: gcr.io/google_containers/hyperkube:v1.1.8
    command: ["/hyperkube", "scheduler", "--address=0.0.0.0", "--master=http://apiserver:8080", "--v=2"]
    network_mode: swarm-private
    environment:
      - "affinity:container==apiserver"
  kubelet:
    image: gcr.io/google_containers/hyperkube:v1.1.8
    command: ['/hyperkube', 'kubelet', '--containerized', '--api_servers=http://apiserver:8080', '--v=2', '--address=0.0.0.0', '--enable_server']
    volumes:
      - /://rootfs:ro
      - /sys:/sys:ro
      - /dev:/dev
      - /var/run/docker.sock:/var/run/docker.sock
      - /var/lib/docker/:/var/lib/docker:ro
      - /var/lib/kubelet:/var/lib/kubelet:rw
      - /var/run:/var/run:rw
    privileged: true
    # A kubelet shouldn't run alongside another kubelet - One privileged kubelet per node
    network_mode: swarm-private
    environment:
      - "affinity:container!=*kubelet*"
  proxy:
    image: gcr.io/google_containers/hyperkube:v1.1.8
    command: ['/hyperkube', 'proxy', '--master=http://apiserver:8080', '--v=2']
    privileged: true
    # A proxy should run alongside another kubelet but not alongside another proxy
    network_mode: swarm-private
    environment:
      - "affinity:container==*kubelet*"
```

In this book, I have been stressing on how I prefer using the Docker Compose method to deploy my container apps. This Docker Compose file is a perfect example why. We have seven containers in this compose file. I find it easy as we have all the logic right there in front of us and we have to write minimal code. Now, let's look at the code. The first thing that is different and that we wouldn't have seen so far is that we are declaring `version 2`. This is because version 1.6.2 of Docker Compose has been released (https://github.com/docker/compose/releases/tag/1.6.2). So, to take advantage of the new features, we need to declare that we want to use `version 2`.

The first container we are declaring is interlock. We are going to use interlock as our application router that will make server requests to the Kubernetes API. For this container, we are going to forward ports, `443`, `8080`, and `8443`, to the host. We will then map `/etc/docker` from the host machine to the container. The reason for this is we need the keys to connect to the Swarm API. So, we will take advantage of the bundle we installed earlier in the chapter.

In the command resource, we will tell interlock where to find the certs, the Swarm URL, and lastly, that we want to use `haproxy`. We will then add this container to our overlay network, `swarm-private`. The next thing is in the environment resource, we will set a constraint and we will tell Compose that we can only run interlock on `ucp-03`. We will do this to avoid port collision with the Kubernetes API service. The next container is `etcd`. Not much has changed regarding this since we configured Kubernetes in the scheduler chapter, so we will move on. The next container is the `kubernetes` API service. The one thing we need to call out with this is that we are declaring in the environment resource - `INTERLOCK_DATA={"hostname":"kubernetes","domain":"ucp-demo.local"}`. This is the URL that interlock will look for when it sends the request to the API.

This is the main reason we are running kubernetes to gain the skills of application routing. So, I won't go through the rest of the containers. Kubernetes is very well documented at `http://kubernetes.io/`. I would recommend that you read up on the features and explore Kubernetes—its a beast, there is so much to learn.

So, now we have all our code. Let's run it!

Just to see the end-to-end build process, we will open our terminal and change the directory to the root of our Vagrant repo. If you have the servers built from the earlier topics, issue `vagrant destroy -f && vagrant up`; if not, just a simple `vagrant up` command will do. Once the Puppet run is complete, our terminal should have the following output:

Best Practices for the Real World

We can then log in to our web UI at `https://127.0.0.1:8443`:

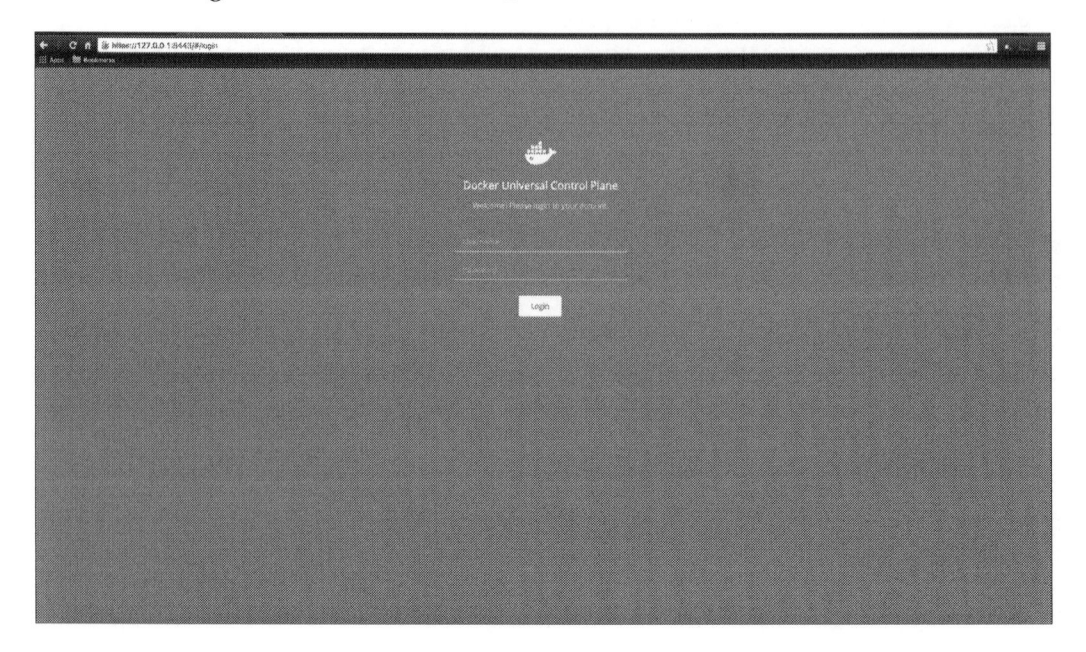

Then, we will log in with the `admin` username and the `orca` password. We should see the following screenshot once we login successfully:

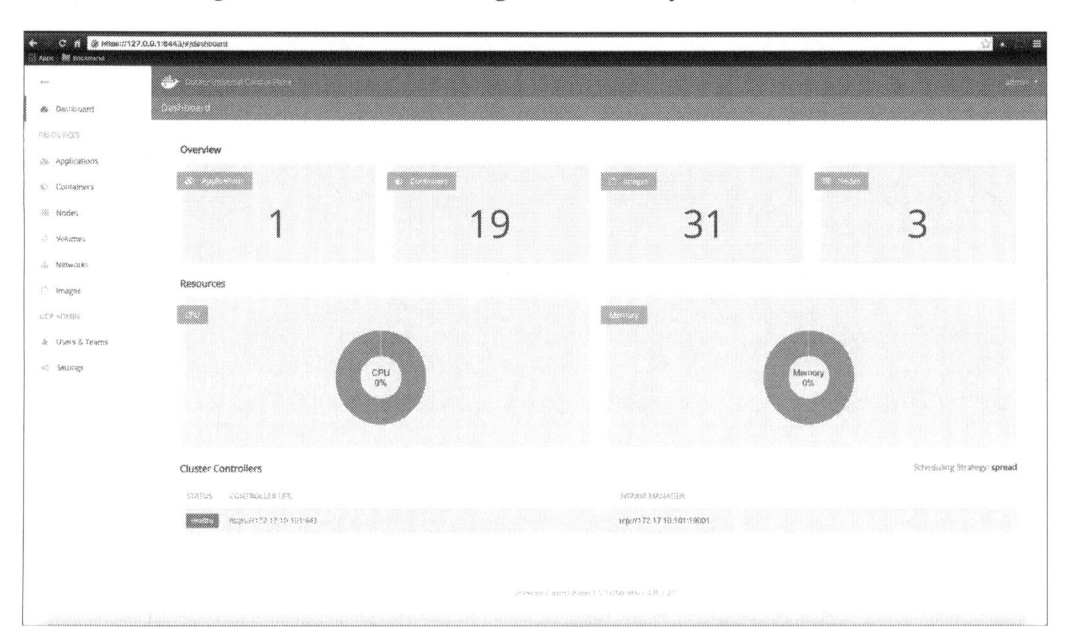

You will note that we have one application now. If we click on the application twice, we can see that Kubernetes is up and running:

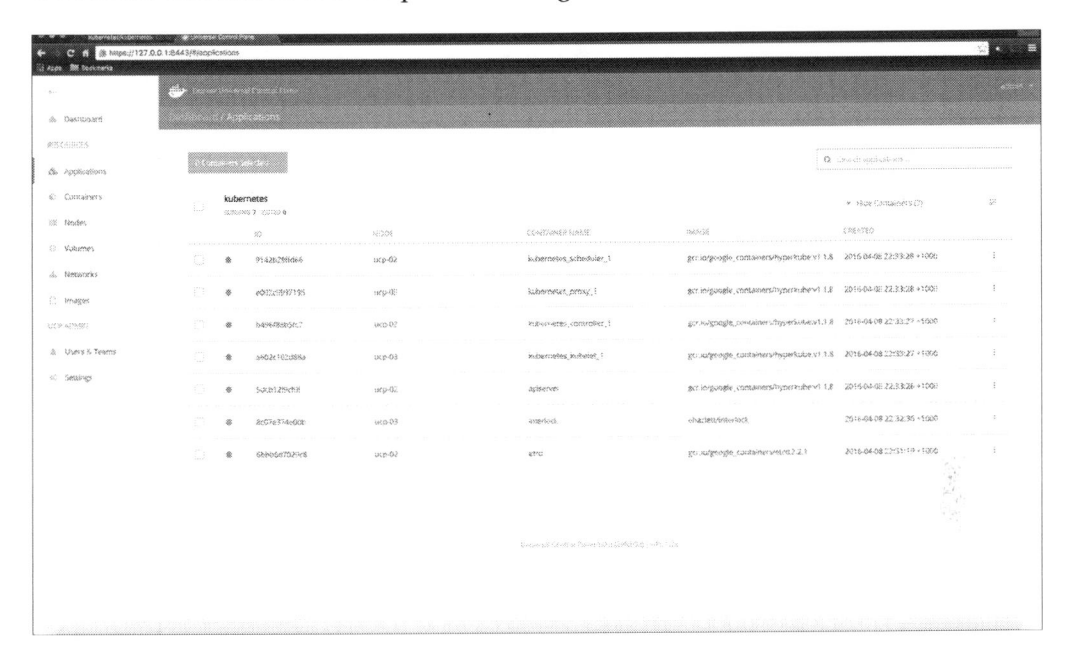

You will note that we have our container split across `ucp-02` and `ucp-03` due to the environment settings in our Docker Compose file. One thing to take note of is that interlock in on `ucp-03` and the Kubernetes API service is on `ucp-02`.

Now that we have built everything successfully, we need to log in to `ucp-03` and download the `kubectl` client. We can achieve that by issuing the following command:

```
$'wget https://storage.googleapis.com/kubernetes-
release/release/v1.1.8/bin/linux/amd64/kubectl'
```

So let's log in to `ucp-03` and issue the `vagrant ssh ucp-03` command from the root of our Vagrant repo. We will then change to root (`sudo -i`). Then, we will issue the `wget` command, as shown in the following screenshot:

We will then make the file executable by issuing the following command:

`$'chmod +x kubectl'`

Now remember that we set a URL in the environment settings for the - `INTERLOCK_DATA={"hostname":"kubernetes","domain":"ucp-demo.local"}` API container. We will need to set that value in our host file. So, use your favorite way to edit files on the local machine, such as vim, nano sed, and so on, and add `172.17.10.103 kubernetes.ucp-demo.local`. The IP address points to `ucp-03` as that is where interlock is running.

Now we are ready to test our cluster. We will do that by issuing the `./kubectl -s kubernetes.ucp-demo.local get nodes` command. We should get the following output after this:

```
root@ucp-03:~# ./kubectl -s kubernetes.ucp-demo.local get nodes
NAME            LABELS                                  STATUS   AGE
a602c102d88a    kubernetes.io/hostname=a602c102d88a     Ready    26m
root@ucp-03:~#
```

As you can see, everything is up and running. If we go back to our UCP console, you can see that the API server is running on `ucp-02` (with the IP address `172.17.10.102`). So, how are we doing this? Interlock is processing the HTTP calls on `8080` and routing them to our API server. This tells us that our application routing is in place. This is a very basic example, but something you should play with as you can really design some slick solutions using interlock.

Summary

In this chapter, we really focused on how to build a Puppet module that is shippable. The use of Hiera and the separation of data from the logic of the module is not only applicable to modules that deploy containers, but for any Puppet modules you write. At some point in your Puppet career, you will either open source a module or contribute to an already open-sourced module. What you have learned in this chapter will be invaluable in both of those use cases. Lastly, we finished with something fun, deploying Kubernetes as the frontend to UCP. In doing this, we also looked at application routing/load balancing. This is obviously a great skill to master as your container environment grows, especially to stay away from issues such as port collision.

Index

A

automated builds
  about 32-35
  Docker Hub, pushing to 35
  in Docker Hub 31

B

Bitbucket
  reference link 73
bitbucket-server
  reference link 66

C

coding
  about 179
  Docker UCP, configuring 180-192
  Kubernetes 192-198
coding, with .erb files
  about 55
  Docker Compose up, with Puppet 57, 58
  module, writing with Docker Compose 55-57
coding, with resource declarations
  about 48
  file structures 49
  module, running 53, 54
  module, writing 49-52
continuous delivery (CD) process 24
Consul
  modules, reference link 122
  reference link 76
  used, for monitoring 168
CoreOS project flannel
  reference link 75
current working directory (cwd) 188

D

Docker
  installing 15
  manifests, setting 15
Docker_bitbucket module, building with docker-compose
  about 70
  coding 71
  running 72, 73
Docker_bitbucket module, building with manifest resources
  coding 64-66
  running 67-70
  skeleton, creating 63
Docker Compose
  about 38-40
  reference link 194
Dockerfiles
  about 36-38
  URL 30, 36
Docker Hub
  about 19
  account, creating 24-26
  account, URL 24
  automated builds 31-35
  Docker Compose 38-40
  Dockerfiles 36, 37
  images, authored by developers 20, 23
  official images 20
  official images, exploring 26-31

overview 20-24
  puppet manifest 41
  pushing to 35
  URL 20
Docker networking
  about 95
  code 95-98
  prerequisites 95
  reference link 95
Docker Swarm
  about 121
  architecture 122
  coding 122-133
  reference link 173
Docker UCP (Universal Control Plane)
  about 121, 133
  architecture 133
  coding 134-143

E

Elasticsearch, Logstash, and Kibana (ELK)
  about 113
  reference link 99
environment
  reference link 178
etcd
  reference link 75

F

Flocker
  reference link 63
Forge
  URL 20

H

Hiera
  about 177
  data components 178
  reference link 104
  tips and tricks 179
hyperkube
  reference link 144

I

interlock
  reference link 172

K

kubectl client
  reference link 148
Kubernetes
  about 143
  architecture 143
  coding 144-153
  reference link 195
  URL 143
Kubernetes pod
  reference link 146

L

logging
  about 155
  code, building 156-162
  code, Logstash 163-166
  solution 156

M

monitoring
  about 167
  Consul, using 168-172

P

parameters, params.pp
  $consul_advertise 50
  $consul_bootstrap_expect 50
  $container_hostname 50
  $docker_image 50
  $docker_tcp_bind 50
  $docker_version 50
pod 146
Puppet Forge
  about 11, 12
  puppetfile, creating 13, 14
  reference link 12
puppet manifest 41

Puppet module generator 44-48
Puppet module skeleton
  building 43

R

recovery techniques
  about 172
  built-in HA 172-176
registrator
  reference link 79

S

service discovery
  about 75
  module 78-94
  theory 75-77
solution design
  about 99
  Consul cluster 100-112
  ELK stack 100
  server setup 102, 103
  summarizing 101
state
  decoupling 62
  versus stateless 62, 63

V

vagrant
  installing 1
  reference link 6
vagrant installation
  about 2
  on Mac OSX 2-8
  Vagrantfile, creating 9-11
vagrant-template
  reference link 9
VirtualBox
  download link 2

W

Weave Works
  reference link 75

Made in the USA
San Bernardino, CA
25 March 2017